General Physics

General Physics

Nelson Bolívar

www.arclerpress.com

General Physics

Nelson Bolívar

Arcler Press

224 Shoreacres Road

Burlington, ON L7L 2H2

Canada

www.arclerpress.com

Email: orders@arclereducation.com

Arcler Press publishes wide variety of books and eBooks. For more information about Arcler Press and its products, visit our website at www.arclerpress.com

ABOUT THE AUTHOR

Nelson Bolivar completed his PhD. in physics from the University of Lorraine, France in 2014 and also from the Central University of Venezuela (U.C.V.). His expertise is in condensed matter and applications in quantum field theories. His interest areas are in spintronic devices and correspondences between general relativity and condensed matter as a new modeling method. He is currently an associated professor at the Central University of Venezuela.

TABLE OF CONTENTS

LIST OF FIGURES

LIST OF TABLES

LIST OF ABBREVIATIONS

ASTM   American Society for Testing and Materials
BCC   Body-Centered-Cubic
CFD   Compression Force Deflection
CT   Compact Tension
FCC   Face-Centered-Cubic
HCP   Hexagonal Close Packed
ISO   International Standardization Organization
LEFM   Linear Elastic Fracture Mechanics
NDE   Non-Destructive Examination
POI   Point of Interest
SENB   Single Edge Notch Bend
SI   International System
SiO_2 Silicon Dioxide
STM   Scanning Tunneling Microscope
UTS   Ultimate Tensile Strength

PREFACE

Physics is regarded as the fundamental physical science. However, in recent times, natural philosophy and physics were interchangeably used for a branch of science that aims at the discovery and creation of the basic laws of nature. With the developments in modern science, Physics played an important role in defining the laws of nature which cannot be explained using other branches of science such as astronomy, geology, engineering, and chemistry. Physics also plays an essential role in all the scientific disciplines including astrophysics, biophysics, psychophysics, and geophysics.

A vast variety of topics are discussed, among which includes Newton laws of motion, mechanics, statistical mechanics, waves, thermodynamics, etc. Most of the physical laws of nature are demonstrated in a simple mathematical language. The readers are expected to have minimal experience in fundamental concepts of mathematics in order to completely understand the physical relations.

Physics is considered the science of experiments which is fundamentally based on experimental methods. Physics was discovered by Galileo Galilei in the 17th century. He was of the opinion that it is indispensable for us to simplify the working conditions of a phenomenon to completely understand it. The process of understanding is not instantaneous; rather it operates by trials and mistakes taking place in a series of experimental events. The understanding process can ultimately result in the formulation of physical laws after passing through different intermediate steps.

Chapter 1 introduces the readers with fundamental concepts of physics as a fundamental science. A brief comparison between classical and quantum physics is also presented. Chapter 2 focuses on Newton laws of motion and their impact on everyday life. Chapter 3 briefly discusses the concepts of space, time, and motion in physical terms. Moreover, the detailed description about the measurement of different physical quantities is also provided in Chapter 3.

Chapter 4 introduces us with essential of engineering mechanics and laws governing the mechanics of materials. Chapter 5 focuses on the fundamental concepts of work, energy, and power with sufficient emphasis on the applications of these concepts in everyday life.

Physics also plays a vital role in understanding the science of materials. Chapter 6 briefly illustrates the structure of different types of materials. Chapter 7 discusses the mechanical properties of materials and their performance in light of physical laws. Finally, Chapter 8 discusses different aspects of energy transfer in the form of heat. Heat transfer by conduction, convection, and radiation phenomena are briefly explained.

The book is fundamentally self-contained which is equally suitable to be treated as a reference book, course book or self-study material. Readers from diverse backgrounds (including students, professionals, and professors) can benefit from this book on fundamentals of Physics.

—*Author*

Chapter
1

Fundamentals of Physics

CONTENTS

1.1. INTRODUCTION

The modern physics is built on three essential theories which include the theory of relativity, quantum mechanics and statistical mechanics. Each of these theories has focused on the idea that the characteristics of the physical world may be far dissimilar from what we perceive and give for granted (Yu and Cardona, 1996).

The theory of special relativity tells us that nothing can ever travel faster than the speed of light. It points out to the fact that there is no absolute time nor is space 'like a uniformly flowing river' (Macdonald and Johnson, 2005). The concept of 'the present' or 'now' is not absolute. It is something that everybody can agree on-every person has his/her own reserved 'now' and each of them is related by a set of specific transformations, space and time are no longer separated but form a unique entity, spacetime. On the other hand, the theory of general relativity concludes that spacetime is curved; and that the universe might be expanding from the initial singularity (i.e., the big bang), and will probably continue to expand until the universe, ubiquitously, is uniformly dark and cold (Reimann and Hänggi, 2002; Rotkin and Subramoney, 2006).

Statistical mechanics or statistical thermodynamics provides the fundamental concept of entropy and the second law of thermodynamics, i.e., the entropy of the closed systems never decreases. Thermodynamic laws tell us about the direction in which takes place a chemical or physical reaction. However, the laws of physics make no distinction between temporal directions, i.e., 'into the past' or 'into the future' the thermodynamics forbids processes that could violate its laws. If we observe something taking place 'the wrong way round' it appears very abnormal indeed, e.g., we often witness breaking of eggs but never see them to reassemble themselves. The logical unidirectionality of different events outlines an 'arrow of time' which is congruent with the laws of thermodynamics.

Statistical mechanics tries to elucidate the characteristics of matter in bulk state in terms of the collective behavior of a large number of constituents. Statistical mechanics tells us that entropy is the measure of a system's degree of disorder which can be possessed by a physical system. Entropy imposes that the natural direction for a system to evolve is in the direction where entropy never decreases. Amongst various other things, it is probable that the universe, with time, could evolve into a condition of maximum disorder (entropy) in which the universe becomes a cold, amorphous, uniform blob, i.e., the so-called thermal death of the universe (Reimann and Hänggi, 2002; Schmid and Ziegelmann, 2017).

Quantum mechanics is a whole new world of physics which introduces us to non-traditional laws of physics. It seems that quantum mechanics provides to us a universal opinion in which:

- There exists a loss of certainty and an unremovable, unavoidable randomness encompasses the physical world. Albert Einstein was extremely dissatisfied with this quantum physical concept, as stated in his well-known saying: "God does not play dice with the universe." It is also predicted that the method of performing an observation may disturb the focused subject in an irrepressibly random way (even if there is no physical contact of the object with the outside world).

- All the physical systems seem to behave as if they are performing a variety of mutually exclusive tasks simultaneously. For example, an electron bombarded at a wall having two holes in it can seem to act as if it passes through both of the holes simultaneously (Popper, 1950; Kofler and Brukner, 2007).

- Broadly disjointed physical systems can act as if they are somehow entangled by a 'spooky action at a distance.' In this way, these physical systems are interconnected in mysterious ways that seem to defy either the rules of special relativity or the laws of probability (Brandt and Physics, 1973; Boyer, 1984).

The third property of quantum mechanics mentioned above leads us to the deduction that there are some facets of the contemporary physical world which are hard to be termed objectively 'real' (Averill and Keating, 1981; Schulze et al., 2000). In short, all of the above three points clearly defy our classical standpoint of the world.

1.2. CLASSICAL PHYSICS

Before the introduction of quantum mechanics for understanding the affairs of the physical world, it is worthwhile to comprehend the classical laws of physics. According to the concepts of classical physics (i.e., pre-quantum physics), it is essentially believed that there exists an 'objectively real world' which continues to exist indefinitely irrespective of our existence (Misner and Wheeler, 1957; Bergman, 1978). These classical physics concepts are not associated with any individual person rather it seems to be the standpoint of Galileo Laplace, Einstein, Newton, and numerous other scientists. These classical concepts seem to reflect an intuitive comprehension of reality. This concept of classical physics can be called 'objective reality' (Figure 1.1).

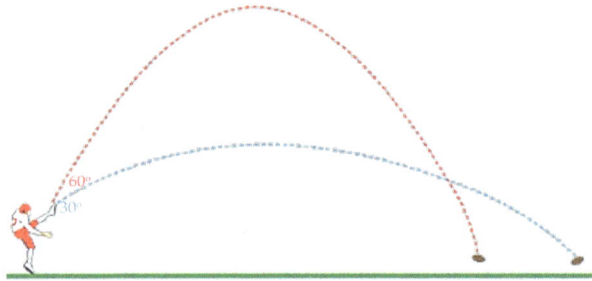

Figure 1.1. Physics of a football travel. The angle of the kick helps govern the travel distance.

Source: https://entertainment.howstuffworks.com/physics-of-football1.htm.

The equations pertaining to the theories of classical physics are included in Newtonian mechanics, Einstein's theory of general relativity, and Maxwell's theory of the electromagnetic field. These equations are believed to demonstrate what is 'really happening' in a physical system (Arnold, 1994; Blandford and Thorne, 2008). For instance, it is presumed that every particulate entity possesses a definite velocity and position and that the explanation of Newton's equations for a moving particle is a perfect illustration of what a particle is 'really doing' (Figure 1.2).

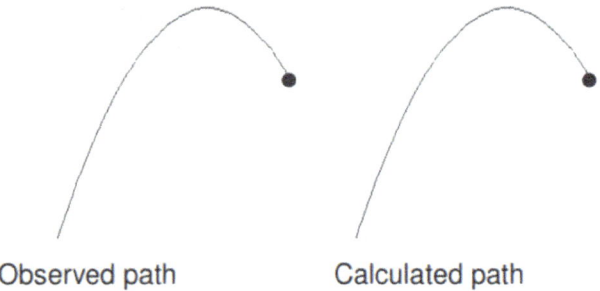

Observed path   Calculated path

Figure 1.2. Comparison of observed and calculated tracks of a tennis ball according to the concepts of classical physics.

Source: http://physics.mq.edu.au/~jcresser/Phys301/Chapters/Chapter1.pdf.

Within the above-mentioned view of reality, a particle in motion through space can be considered like a tennis ball floating in the air, as if it possesses, at any given time, a definite velocity and position (Wilcox, 1966; Da Costa and Doria, 1991). Moreover, it would continue to possess that definite velocity and position irrespective of the fact that anyone is observing its behavior. That is the typical and classical way of observing things (Popper, 1950; Greengard, 1994). It is, however, up to us to choose whether or not we need to measure the pre-existing velocity and position. They both possess some definite values at a certain occasion, but it is entirely a function of our trial inventiveness to measure these values with an appropriate level of precision. There is an implied conviction that by improving our experiments—e.g., by quantifying to the 10th decimal place followed by the 100th and then the 1000th—we can approach to the actual/real values of the velocity and position of the particle (Yao, 2003; Holstein and Donoghue, 2004). There is no physical law in classical physics which clearly defies the fact that we can determine these values of velocity and position to the desired decimal places. However, the only constraint is our experimental ingenuity. Principally, it is also possible to calculate, with ultimate accuracy, the impending behavior of any existing physical system by solving Maxwell's and Newton's equations. Practically, there are certain limits to accuracy in measurement and calculation, but principally no such limits exist (Gurtin and Martins, 1976; Anandan, 1988).

1.3. CLASSICAL RANDOMNESS AND IGNORANCE OF INFORMATION

For a macroscopic object, we realize that it is not possible to measure the velocities and positions of all the particles constituting the object. For example, a bottle full of air at room temperature roughly contains 10^{26} particles rushing around in the bottle, colliding each other and with the bottle walls (Dürr et al., 1992; Berta et al., 2012, 2013). There is absolutely no way to measure the velocities and positions of all the gas particles at a certain instance. However, we can safely believe that every particle possesses a definite velocity and position at each instant, but it is extremely difficult to extract all the information (D'Ariano et al., 2005; Farenick et al., 2011).

Similarly, we cannot predict the movement of a pollen grain suspended in a certain liquid. The pollen grains are in constant Brownian motion (i.e., random walk) due to collisions with liquid molecules. According to the laws of classical physics, the relevant information is 'really present there' but we just cannot retrieve it (Figure 1.3).

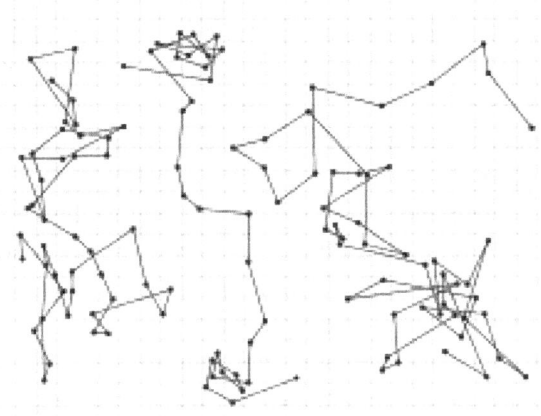

Figure 1.3. Brownian motion exhibited by pollen grains suspended in a liquid.

Source: https://en.wikipedia.org/wiki/Brownian_motion.

Random behavior (i.e., Brownian motion) only *seems* random because we lack the necessary information to demonstrate it exactly. It is not categorically random because we can get the same results if we repeat an experiment under *exactly the* same conditions, and henceforth, the consequence of the experiment would be impeccably predictable (Dodis and Renner, 2006; Renner, 2008).

In the case of gases, we tend to accept a certain ignorance level about the possible data that we could obtain. Because of this ignorance, we cannot make accurate estimates about the future performance of the gas. We can compensate for the flaws of the ignorance by employing statistical techniques to estimate the likelihoods of the gas particles acting in several possible ways. For example, it is probable to demonstrate that the odds of all the gas particles instinctively rushing towards a particular end of the bottle is roughly 1 in 10^{26}, i.e., appallingly unlikely (Calude et al., 2006; Bailly and Longo, 2007; Longo et al., 2011).

1.4. QUANTUM PHYSICS

The classical view of the world can be applied conveniently at the everyday (i.e., macroscopic) level as most of the modern engineering issues rely on this. However, there are certain phenomena at the macroscopic level which are impossible to grasp using the concepts of classical physics which include the existence of solid objects, the color of a heated object, etc. (Piron, 1976; Economou, 1983; Eisberg and Resnick, 1985).

On the other hand, microscopic systems (i.e., atoms and molecules) usually exhibit non-classical behavior but it is essentially present at all scales (Sudbery, 1986; French and Ladyman, 2003; El Naschie, 2004). The types of behaviors exhibited by microscopic systems defying classical physics are expressed below:

- Intrinsic randomness;
- Interference phenomena (e.g., particles behaving as waves);
- Entanglement.

1.4.1. Intrinsic Randomness

It is not possible to construct *any* physical system which possesses all of the precisely specified physical attributes at the same instance—e.g., it is impossible to precisely determine both the momentum and position of a particle at the same instant of time (Cohen-Tannoudji et al., 1997; Buks and Roukes, 2002). For example, if we confine a particle in a small box, thereby providing us a precise indication of its position, and try to measure its velocity at any moment, we find that the velocity of the particular particle always fluctuates in a randomly from one measurement to the next (Goldstein et al., 1981; Yuan et al., 2015). For example, for an electron confined in a box of 1 μm in size, the velocity can be estimated to fluctuate by at least ±50 m/s. It is impossible to reduce this randomness despite the refinement of the experiment. The randomness can never be eliminated, and building the tinier box just renders the situation worse (Nomura et al., 2004; Nomura, 2012).

More commonly, it is established fact that for any experiment performed multiple times under exactly the same conditions there will be some physical quantity or property of the system constituting the experiment, which will always produce randomly fluctuating results from one round of the experiment to the next. This phenomenon does not imply that our experimental conditions and procedure is faulty rather the randomness is irreducible, i.e., it cannot be entirely eliminated by improving the experimental technique (Courbage and Prigogine, 1983; Hayashi, 2008).

The above-mentioned concept is informing us about the fundamental reality of the universe that there is always a limit to the extent of information we can extract from a physical system. Apparently, we cannot know about a system with as much precision as we assumed we could keep in mind the classical physics (Nicolis and Baras, 1987; Miyaguchi and Akimoto, 2011). This fact tempts the scientists to ponder on the fact that the missing information still exists but it is simply inaccessible for some reason. For example, a particle with a known position also possesses a precise velocity (or momentum), but we merely cannot measure the value of this momentum. Thus, the randomness seems to exist in such situations is not an outcome of our ignorance (Bloch, 2010; Dhara et al., 2014).

1.4.2. Interference

Most of the microscopic physical systems behave in such a way that they are performing mutually exclusive events at the same time. The best example of such physical behaviors is the well-known two-slit experiment which involves the firing of one electron at a time but we observe two narrow slits on the screen (Khrennikov, 2005; Scarani, 2006). The electrons seem to hit the observation screen positioned beyond the slit-screen. It is expected that the electrons will hit the second screen in positions immediately opposite to the two slits of the first screen. It is also observed that the electrons reaching the observation screen are likely to arrive at preferred sites which exhibit the features of an interference pattern produced by waves, i.e., the pattern by electronic interference will resemble the pattern formed by light waves headed towards the slits (Rauch, 1993; Gerlich et al., 2011) (Figure 1.4).

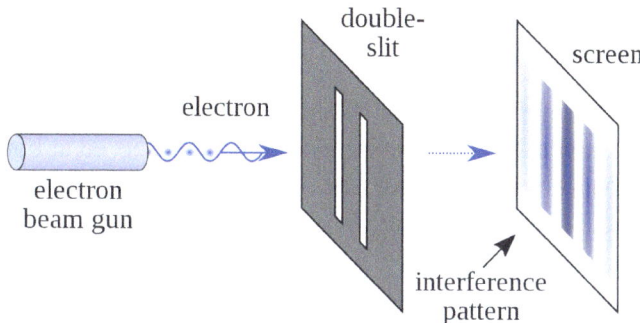

Figure 1.4. Schematic illustration of interference phenomenon.

Source: https://en.wikipedia.org/wiki/Double-slit_experiment.

The comprehensive characteristics of the interference pattern are determined by the extent of slit separation, i.e., increasing the separation distance results in a finer interference pattern. This phenomenon suggests that an electron (being a particle) that has the ability to pass through a single slit, somehow, manages to have the 'knowledge' of the second slit's position. In the case of the electron not having the knowledge about the position of both slits, the pattern formed on the second screen would appear to have no interference. This is proved by the fact that the electrons hit the screen at two sites directly opposite the position of slits (Wheeler, 1985; Zeilinger, 1999). The above-mentioned phenomenon explains the wave nature of the electrons which can pass through two holes/slits at the same time and ultimately form an interference pattern (Boyarsky and Góra, 1992; Nairz et al., 2003).

This tendency for a quantum system to act as if they can exist at two places at a single moment, or more commonly in various states at the same instance of time is known as 'the superposition of states.' This superposition of states is a singular but primary characteristic of quantum systems which results in the creation of a mathematical explanation based on the concepts of vector linear spaces (Wernsdorfer and Sessoli, 1999; Sinha et al., 2010).

1.4.3. Entanglement

Entanglement is a quantum phenomenon which provides an explanation for the connection of two or more systems with each other. For instance, we buy a pair of shoes and keep one shoe in one room of our office while the other in our home. Somehow, we can discover that the left shoe is in the office while the right one is in the house. But this does not change the fact that both shoes are connected to each other and complement each other. If this strange custom of splitting up flawlessly good pairs of shoes and distributing one at random to the office and the other to the house were repeated several times, it is, nonetheless, always the case that the outcomes observed at both places were always impeccably correlated, i.e., a right shoe paired off with a left shoe (Ursin et al., 2007; Riedel et al., 2010) (Figure 1.5).

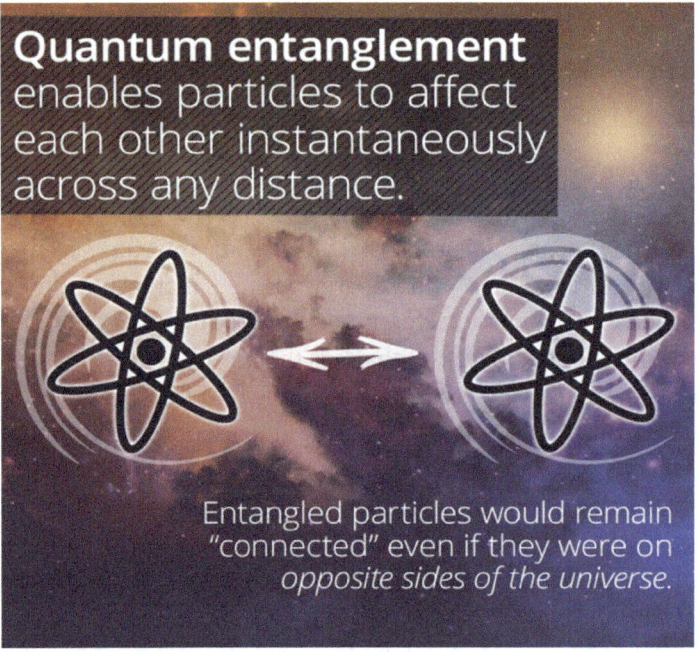

Figure 1.5. Entanglement of particles is governed by physical law of entanglement.

Source: https://genesisnanotech.wordpress.com/2018/04/26/einsteins-theory-of-entanglement-aka-spooky-action-goes-massive/.

Similar experimental ventures can be performed with atomic particles as the spins of particle pairs are paired off. It is an established fact that spins of the particles are in exactly opposed to each other in such a way that the total angular momentum is zero (Vedral et al., 2003; Vidal et al., 2003). In the case of the shoes, the measurement of spin is made when the particle (i.e., shoe) arrives in office or in house. On the contrary, in the case of atomic particles, the measurement of spins of particles is no straightforward, i.e., it goes outside the domains of simple examples of right or left shoe, but the concept is, nonetheless, to measure the correlations existing between the particles' spins. Hence, considering quantum physics, it is quite probable for the spinning particles to be formed in conditions for which the association between the measured values of spin is *greater* than the values of correlation existing in classical physics. These systems are known to be existing in 'entangled states,' i.e., the quantum states with classical analogs. Amongst many other facts, this entanglement phenomenon suggests that the two entangled systems are in constant and instantaneous 'communication' with each other which is quicker than the speed of light, though, this concept contradicts special theory of relativity. However, this phenomenon can take place at atomic or subatomic states but we cannot precisely measure the position and velocity of the particle simultaneously to determine the exact nature of this phenomenon (Zheng and Guo, 2000; Jozsa and Linden, 2003).

1.5. OBSERVATION, DATA, AND THE PRINCIPLES OF PHYSICS

Most of the modern physical theories did not come into existence merely by pure deliberation (except, for instance, general relativity). The common characteristic of all physical theories is that these theories deal with the data about physical systems which we obtain through observation or experimentation. For example, equations of Maxwell for the EM field are not just a concise summary of the observed properties of magnetic and electric fields and any related currents and charges (Rhoads and Thorn, 1996; Lau et al., 2004). These equations have been abstracted from the outcomes of countless experiments performed over 100 years, as well as with some smart interpolation on the part of Maxwell. Likewise, comments can be made regarding Newton's laws of motion (D'Urso and Calera Belmonte, 2006; Lv et al., 2015). We collect data either by casual reflection or through controlled experiment, for example, on the temperature, on the motion of physical objects, on the pressure, volume of liquids, solids, or gases and so on. Through observation of this data, regularities are observed and summarized aptly using the following equations:

$$F = ma \qquad \text{— Newton's 2nd law;}$$

$$\nabla \times E = -\frac{\partial B}{\partial t} \qquad \text{— Maxwell's equations (Faraday's law);}$$

$$PV = NkT \quad \text{— Ideal gas law (this is not a fundamental law)}$$

These equations represent relationships among *information* acquired through *observation* of numerous physical systems and as such are a brief mode of summarizing the relationship among the information, or the data, gathered about a physical system. The laws are articulated in a way consistent with how we comprehend the world from the point of view of classical physics in that the symbols substitute accurately known values of the physical quantities they characterize (Uchida et al., 2006; Zhang et al., 2017).

These early researchers did not know that the information they were collecting was not amply refined to demonstrate that there were fundamental limits to the precision with which they could measure physical properties. Additionally, there was some information which might have been taken for granted by them (Zhou et al., 2006; Ladak et al., 2010). Certain laws of nature place restrictions on the information which can be gathered about any physical system (Boughn et al., 1977; Brewer et al., 1986). The gathered information was eventually organized into mathematical statements which constitute *classical* laws of physics: Newton's laws of motion or Maxwell's equations. However, in the late 19th century and on into the 20th century, different experimental proof started to accumulate which proposed that there was somewhat seriously incorrect with the classical laws of physics: therefore, the information could no longer be fitted to the equations, or, we can say, the theory was unable to explain the observed experimental results (Pinch, 1985; Goris et al., 2006; Ramesha et al., 2007). There were two choices: either formulate new theories or else modify the existing theories. Former approach was adopted. Eventually, a new set of laws of nature were formulated, the laws of quantum mechanics, which were basically a set of laws regarding the *information* that could be acquired about the physical world (Cha et al., 2007; Ni, 2008).

REFERENCES

1.   Anandan, J., (1988). Geometric angles in quantum and classical physics. *Physics Letters A, 129*(4), 201–207.

2.   Arnold, V. I., (1994). Mathematical problems in classical physics. *Trends and Perspectives in Applied Mathematics* (Vol. 1, pp. 1–20). Springer, New York, NY.

3.   Averill, E., & Keating, B. F., (1981). Does interactionism violate a law of classical physics? *Mind, 90*(357), 102–107.

4.   Bailly, F., & Longo, G., (2007). Randomness and determinism in the interplay between the continuum and the discrete. *Mathematical Structures in Computer Science, 17*(2), 289–305.

5.   Bergman, D. J., (1978). The dielectric constant of a composite material—a problem in classical physics. *Physics Reports, 43*(9), 377–407.

6.   Berta, M., Fawzi, O., & Wehner, S., (2012). Quantum to classical randomness extractors. In: *Annual Cryptology Conference* (Vol. 1, pp. 776–793). Springer, Berlin, Heidelberg.

7.   Berta, M., Fawzi, O., & Wehner, S., (2013). Quantum to classical randomness extractors. *IEEE Transactions on Information Theory*, 1168–1192.

8.   Blandford, R. D., & Thorne, K. S., (2008). Applications of classical physics. *Lecture Notes, California Institute of Technology, 12*(1), 1–10.

9.   Bloch, M., (2010). Channel intrinsic randomness. In: *2010 IEEE International Symposium on Information Theory* (Vol. 1, pp. 2607–2611). IEEE.

10.   Boughn, S. P., Fairbank, W. M., Giffard, R. P., Hollenhorst, J. N., McAshan, M. S., Paik, H. J., & Taber, R. C., (1977). Observation of mechanical Nyquist noise in a cryogenic gravitational-wave antenna. *Physical Review Letters, 38*(9), 454.

11.   Boyarsky, A., & Góra, P., (1992). A dynamical system model for interference effects and the two-slit experiment of quantum physics. *Physics Letters A, 168*(2), 103–112.

12.   Boyer, T. H., (1984). Derivation of the blackbody radiation spectrum from the equivalence principle in classical physics with classical electromagnetic zero-point radiation. *Physical Review D, 29*(6), 1096.

13.   Brandt, L. W., & Physics, C., (1973). The physics of the physicist and

the physics of the psychologist. *International Journal of Psychology*, *8*(1), 61–72.

14. Brewer, J. H., Kreitzman, S. R., Noakes, D. R., Ansaldo, E. J., Harshman, D. R., & Keitel, R., (1986). Observation of muon-fluorine" hydrogen bonding" in ionic crystals. *Physical Review B*, *33*(11), 7813.

15. Buks, E., & Roukes, M. L., (2002). Quantum physics: Casimir force changes sign. *Nature*, *419*(6903), 119.

16. Calude, C. S., Staiger, L., & Terwijn, S. A., (2006). On partial randomness. *Annals of Pure and Applied Logic*, *138*(1–3), 20–30.

17. Cha, J. W., Yum, S. S., Chang, K. H., & Oh, S. N., (2007). Estimation of the melting layer from a micro rain radar (MRR) data at the cloud physics observation system (CPOS) site at Daegwallyeong Weather Station. *Journal of the Korean Meteorological Society*, *43*(1), 77–85.

18. Cohen-Tannoudji, C., Dupont-Roc, J., & Grynberg, G., (1997). Photons and atoms-introduction to quantum electrodynamics. In: Claude, C. T., Jacques, D. R., & Gilbert, G., (eds.), *Photons and Atoms-Introduction to Quantum Electrodynamics* (Vol. 1, pp. 486). ISBN 0–471-18433-0. Wiley-VCH.

19. Courbage, M., & Prigogine, I., (1983). Intrinsic randomness and intrinsic irreversibility in classical dynamical systems. *Proceedings of the National Academy of Sciences*, *80*(8), 2412–2416.

20. D'Ariano, G. M., Presti, P. L., & Perinotti, P., (2005). Classical randomness in quantum measurements. *Journal of Physics A: Mathematical and General*, *38*(26), 5979.

21. D'Urso, G., & Calera, B. A., (2006). Operative approaches to determine crop water requirements from earth observation data: Methodologies and applications. In: *AIP Conference Proceedings* (Vol. 852, No. 1, pp. 14–25). AIP.

22. Da Costa, N. C., & Doria, F. A., (1991). Classical physics and Penrose's thesis. *Foundations of Physics Letters*, *4*(4), 363–373.

23. Dhara, C., De La Torre, G., & Acín, A., (2014). Can observed randomness be certified to be fully intrinsic? *Physical Review Letters*, *112*(10), 100402.

24. Dodis, Y., & Renner, R., (2006). On the impossibility of extracting classical randomness using a quantum computer. In: *International Colloquium on Automata, Languages, and Programming* (Vol. 1, pp. 204–215). Springer, Berlin, Heidelberg.

25. Dürr, D., Goldstein, S., & Zanghi, N., (1992). Quantum chaos, classical randomness, and Bohmian mechanics. *Journal of Statistical Physics*, *68*(1/2), 259–270.

26. Economou, E. N., (1983). *Green's Functions in Quantum Physics* (Vol. 3, pp. 11–31). New York: Springer.

27. Eisberg, R., & Resnick, R., (1985). Quantum physics of atoms, molecules, solids, nuclei, and particles. In: Robert, E., & Robert, R., (eds.), *Quantum Physics of Atoms, Molecules, Solids, Nuclei, and Particles* (2nd edn., Vol. 2, p. 864). ISBN 0–471–87373-X. Wiley-VCH.

28. El Naschie, M. S., (2004). The concepts of E infinity: An elementary introduction to the Cantorian-fractal theory of quantum physics. *Chaos, Solitons and Fractals*, *22*(2), 495–511.

29. Farenick, D., Plosker, S., & Smith, J., (2011). Classical and nonclassical randomness in quantum measurements. *Journal of Mathematical Physics*, *52*(12), 122204.

30. French, S., & Ladyman, J., (2003). Remodeling structural realism: Quantum physics and the metaphysics of structure. *Synthese*, *136*(1), 31–56.

31. Gerlich, S., Eibenberger, S., Tomandl, M., Nimmrichter, S., Hornberger, K., Fagan, P. J., & Arndt, M., (2011). Quantum interference of large organic molecules. *Nature Communications*, *2*, 263.

32. Goldstein, S., Misra, B., & Courbage, M., (1981). On intrinsic randomness of dynamical systems. *Journal of Statistical Physics*, *25*(1), 111–126.

33. Goris, L., Poruba, A., Hod'Ákova, L., Vaněček, M., Haenen, K., Nesládek, M., & Manca, J. V., (2006). Observation of the sub gap optical absorption in polymer-fullerene blend solar cells. *Applied Physics Letters*, *88*(5), 052113.

34. Greengard, L., (1994). Fast algorithms for classical physics. *Science*, *265*(5174), 909–914.

35. Gurtin, M. E., & Martins, L. C., (1976). Cauchy's theorem in classical physics. *Archive for Rational Mechanics and Analysis*, *60*(4), 305–324.

36. Hayashi, M., (2008). Second-order asymptotic in fixed-length source coding and intrinsic randomness. *IEEE Transactions on Information Theory*, *54*(10), 4619–4637.

37. Holstein, B. R., & Donoghue, J. F., (2004). Classical physics and quantum loops. *Physical Review Letters*, *93*(20), 201602.

38. Jozsa, R., & Linden, N., (2003). On the role of entanglement in quantum-computational speed-up. *Proceedings of the Royal Society of London: Series A: Mathematical, Physical and Engineering Sciences, 459*(2036), 2011–2032.

39. Khrennikov, A., (2005). The principle of supplementarity: A contextual probabilistic viewpoint to complementarity, the interference of probabilities and incompatibility of variables in quantum mechanics. *Foundations of Physics, 35*(10), 1655–1693.

40. Kofler, J., & Brukner, Č., (2007). Classical world arising out of quantum physics under the restriction of coarse-grained measurements. *Physical Review Letters, 99*(18), 180403.

41. Ladak, S., Read, D. E., Perkins, G. K., Cohen, L. F., & Branford, W. R., (2010). Direct observation of magnetic monopole defects in an artificial spin-ice system. *Nature Physics, 6*(5), 359.

42. Lau, C. N., Stewart, D. R., Williams, R. S., & Bockrath, M., (2004). Direct observation of nanoscale switching centers in metal/molecule/metal structures. *Nano Letters, 4*(4), 569–572.

43. Longo, G., Palamidessi, C., & Paul, T., (2011). Some bridging results and challenges in classical, quantum and computational randomness. *Randomness Through Computation: Some Answers, More Questions, 1*(1), 73–91.

44. Lv, B. Q., Xu, N., Weng, H. M., Ma, J. Z., Richard, P., Huang, X. C., & Strocov, V. N., (2015). Observation of Weyl nodes in TaAs. *Nature Physics, 11*(9), 724.

45. Macdonald, J. R., & Johnson, W. B., (2005). Fundamentals of impedance spectroscopy. *Impedance Spectroscopy: Theory, Experiment, and Applications, 1*, 1–26.

46. Misner, C. W., & Wheeler, J. A., (1957). Classical physics as geometry. *Annals of Physics, 2*(6), 525–603.

47. Miyaguchi, T., & Akimoto, T., (2011). Intrinsic randomness of transport coefficient in sub diffusion with static disorder. *Physical Review E, 83*(3), 031926.

48. Nairz, O., Arndt, M., & Zeilinger, A., (2003). Quantum interference experiments with large molecules. *American Journal of Physics, 71*(4), 319–325.

49. Ni, W. T., (2008). From equivalence principles to cosmology: Cosmic polarization rotation, CMB observation, neutrino number asymmetry, Lorentz invariance, and CPT. *Progress of Theoretical Physics Supplement*, *172*, 49–60.

50. Nicolis, G., & Baras, F., (1987). Intrinsic randomness and spontaneous symmetry-breaking in explosive systems. *Journal of Statistical Physics*, *48*(5/6), 1071–1090.

51. Nomura, K., Kamiya, T., Ohta, H., Ueda, K., Hirano, M., & Hosono, H., (2004). Carrier transport in transparent oxide semiconductor with intrinsic structural randomness probed using single-crystalline InGaO 3 (Z nO) 5 films. *Applied Physics Letters*, *85*(11), 1993–1995.

52. Nomura, R., (2012). Second-order resolvability, intrinsic randomness, and fixed-length source coding for mixed sources: Information spectrum approach. *IEEE Transactions on Information Theory*, *59*(1), 1–16.

53. Pinch, T., (1985). Towards an analysis of scientific observation: The externality and evidential significance of observational reports in physics. *Social Studies of Science*, *15*(1), 3–36.

54. Piron, C., (1976). On the foundations of quantum physics. In: *Quantum Mechanics, Determinism, Causality, and Particles* (Vol. 1, pp. 105–116). Springer, Dordrecht.

55. Popper, K. R., (1950). Indeterminism in quantum physics and in classical physics: Part I. *The British Journal for the Philosophy of Science*, *1*(2/3), 117–195.

56. Ramesha, K., Llobet, A., Proffen, T., Serrao, C. R., & Rao, C. N. R., (2007). Observation of local non-centrosymmetry in weakly biferroic $YCrO_3$. *Journal of Physics: Condensed Matter*, *19*(10), 102202.

57. Rauch, H., (1993). Phase space coupling in interference and EPR experiments. *Physics Letters A*, *173*(3), 240–242.

58. Reimann, P., & Hänggi, P., (2002). Introduction to the physics of Brownian motors. *Applied Physics A*, *75*(2), 169–178.

59. Renner, R., (2008). Extracting classical randomness in a quantum world. In *2008 IEEE Information Theory Workshop* (Vol. 1, pp. 360–363). IEEE.

60. Rhoads, B. L., & Thorn, C. E., (1996). Observation in geomorphology. *The Scientific Nature of Geomorphology*, *1*(1), 21–56.

61. Riedel, M. F., Böhi, P., Li, Y., Hänsch, T. W., Sinatra, A., & Treutlein, P., (2010). Atom-chip-based generation of entanglement for quantum metrology. *Nature, 464*(7292), 1170.

62. Rotkin, S. V., & Subramoney, S., (2006). *Applied Physics of Carbon Nanotubes: Fundamentals of Theory, Optics and Transport Devices* (Vol. 1, pp. 1–16). Springer Science and Business Media.

63. Scarani, V., (2006). *Quantum Physics: A First Encounter: Interference, Entanglement, and Reality* (Vol. 1, pp. 7–26). OUP Oxford.

64. Schmid, E. W., & Ziegelmann, H., (2017). *The Quantum Mechanical Three-Body Problem: Vieweg Tracts in Pure and Applied Physics* (Vol. 2, p. 20). Elsevier.

65. Schulze, K. G., Shelby, R. N., Treacy, D. J., Wintersgill, M. C., Vanlehn, K., & Gertner, A., (2000). Andes: An intelligent tutor for classical physics. *Journal of Electronic Publishing, 6*(1), 1–20.

66. Sinha, U., Couteau, C., Jennewein, T., Laflamme, R., & Weihs, G., (2010). Ruling out multi-order interference in quantum mechanics. *Science, 329*(5990), 418–421.

67. Sudbery, A., (1986). *Quantum Mechanics and the Particles of Nature* (Vol. 136, pp. 11–30). Cambridge: Cambridge University Press.

68. Uchida, M., Onose, Y., Matsui, Y., & Tokura, Y., (2006). Real-space observation of helical spin order. *Science, 311*(5759), 359–361.

69. Ursin, R., Tiefenbacher, F., Schmitt-Manderbach, T., Weier, H., Scheidl, T., Lindenthal, M., & Ömer, B., (2007). Entanglement-based quantum communication over 144 km. *Nature Physics, 3*(7), 481.

70. Vedral, V., (2003). Classical correlations and entanglement in quantum measurements. *Physical Review Letters, 90*(5), 050401.

71. Vidal, G., Latorre, J. I., Rico, E., & Kitaev, A., (2003). Entanglement in quantum critical phenomena. *Physical Review Letters, 90*(22), 227902.

72. Wernsdorfer, W., & Sessoli, R., (1999). Quantum phase interference and parity effects in magnetic molecular clusters. *Science, 284*(5411), 133–135.

73. Wheeler, J. A., (1985). Franck-Condon effect and squeezed-state physics as double-source interference phenomena. *Letters in Mathematical Physics, 10*(2/3), 201–206.

74. Wilcox, C. H., (1966). Wave operators and asymptotic solutions of

wave propagation problems of classical physics. *Archive for Rational Mechanics and Analysis*, *22*(1), 37–76.

75. Yao, A. C. C., (2003). Classical physics and the Church-Turing thesis. *Journal of the ACM (JACM)*, *50*(1), 100–105.

76. Yu, P. Y., & Cardona, M., (1996). *Fundamentals of Semiconductors: Physics and Materials Properties* (Vol. 1, pp. 1–26). Springer.

77. Yuan, X., Zhou, H., Cao, Z., & Ma, X., (2015). Intrinsic randomness as a measure of quantum coherence. *Physical Review A*, *92*(2), 022124.

78. Zeilinger, A., (1999). Experiment and the foundations of quantum physics. In: *More Things in Heaven and Earth* (Vol. 1, pp. 482–498). Springer, New York, NY.

79. Zhang, J., Hess, P. W., Kyprianidis, A., Becker, P., Lee, A., Smith, J., & Yao, N. Y., (2017). Observation of a discrete-time crystal. *Nature*, *543*(7644), 217.

80. Zheng, S. B., & Guo, G. C., (2000). Efficient scheme for two-atom entanglement and quantum information processing in cavity QED. *Physical Review Letters*, *85*(11), 2392.

81. Zhou, S. Y., Gweon, G. H., Graf, J., Fedorov, A. V., Spataru, C. D., Diehl, R. D., & Lanzara, A., (2006). First direct observation of Dirac fermions in graphite. *Nature Physics*, *2*(9), 595.

Chapter 2

Forces and Newton's Laws

CONTENTS

2.1. INTRODUCTION

All of us have experienced numerous examples of forces. We are well-known with a muscular force producing a pull or a push. There are a lot of other examples for instance: the Earth's gravitational force; one solid object acting on another object by means of physical contact; the attraction among a plastic rod which a cat has rubbed with her fur; the magnetic attraction for an object made of iron; and the force of a stressed spring. One consequence of a force is to change the current state of motion of a body (Felten, 1984; Thornton and Sokoloff, 1998; Ogborn and Taylor, 2005).

Our focus in this chapter will be on forces. Acceleration, projectile motion, accelerated motion in a uniform manner, and the other types of motion need force (Minstrell, 1984; Itza-Ortiz et al., 2003; Handhika et al., 2016). In order to determine these different motions, we have to first determine the many forces which act upon a body. The study of forces and the allied changes in motion is known as *dynamics*. The conditions leading to equilibrium are of particular interest (White, 1984; Norbury, 1998; Pugh, 2004).

2.2. NEWTON'S FIRST AND THIRD LAWS OF MOTION

This chapter introduces the mental constructs invented by Sir Isaac Newton in the 17th century. Newton's model attempts to explain the relationship between different forces acting on an object and the resultant motion of that object. This model is known as Newtonian mechanics; it does not provide accurate results in all cases (Galili and Tseitlin, 2003; Pfister, 2004). For instance, Einstein's special theory of relativity expands the mechanics for objects traveling at speeds close to the speed of light (which is 3.00×10^8 m/sec). In the demesne of distances, the order of atomic radius (which is 10^{-10} m) quantum theory should be employed to provide an accurate picture of physical phenomena. Nonetheless, in daily experiences, Newton's mechanics offers us a good model in order to understand forces and motion (Anderson, 1990; Roberson et al., 2004).

Let us reflect on the below-mentioned experiences we all probably have had in our lives:

- If a person is riding in an auto and the brakes are applied unexpectedly, he/she experiences an inclination to knock into the windshield.
- A book has been placed on a sheet of paper on a tabletop. Even

after the passage of several hours, the book is still resting on that sheet of paper on the top of the table. Reason? Newton's first law of motion might give us some understanding of alike experiences.

• If a person is riding a sprinting horse and the horse abruptly halts, the person might find himself/herself on the ground in front of him. Why is that so?

Everybody will continue to be in its state of rest or in uniform linear motion unless and until an unbalanced force acts upon it.

The first law of motion is occasionally called the *inertial law*. Recalling the perception of inertia, which is that any object has an inclination to repel change. In the field of mechanics, this property of any object to repel a change in its state of motion is known as its *inertia* and is calculated by the mass of the object (Santavy, 1986; Tinto, 2013; Hecht, 2015).

Newton's first law of motion aids to unite our experiences with resting objects not unexpectedly moving as well as with moving objects facing exertion in stopping them at once. The degree of the slowness of variation in the motion of an object is known as its mass (Terry and Jones, 1986; Kurniawan et al., 2016). While wrestling, two persons hold one another's hands and push with their maximum force and for just a moment, they get locked in a motionless tussle. Each person is pushing hard against his rival. The rival is also exerting an equal force against the first person. According to Newton's third law of motion, it is just the same as if we would be pushing against a brick wall (Figure 2.1). Our force against the brick wall gets balanced by the push of wall against us, similar to how our push was countered by that of our wrestling rival (Maloney, 1984; Brown, 1989).

Figure 2.1. Two men tussling with each other will balance each other's force at any moment. This occurrence is just like a person pushing against the wall, that is, action and reaction are equal in magnitude however opposite in direction.

Source: http://physics.doane.edu/hpp/Resources/Fuller3/pdf/F3Chapter_4.pdf.

Let us consider the example of our physics book which is resting on the study table. The physics book is exerting a downward force on the study table; whereas, the study table exerts a force in the upward direction on the book. This leaves a pair of forces equal in magnitude and have opposite direction. One force is with which the book acts on the desk, and the other force is exerted by the desk on the book (Hellingman, 1992; Savinainen et al., 2005). One force is known as the *action force;* whereas, the other force is known as the *reaction force*. These two forces form a pair of action-reaction forces. Newton's third law of motion unifies such experiences in a simple statement:

For every action, there is an equal and opposite amount of reaction.

Consider an example of a rope tied around a tree. Now a force is exerted on the rope by us as a reaction of which the rope also exerts a force on us. This is an action-reaction pair f forces. There is another action-reaction pair at the other end of the rope as the rope exerts a force on the tree and the tree exerts a force on the rope. If there are two interacting bodies, namely, A and B, the action-reaction pair of forces can be defined by A acting on B and vice-versa, B acting upon A (Cohen, 1987; MaKinster et al., 2002).

Suppose a tug of war competition among two sides A and B, as displayed in Figure 2.2. The rope is exerting a force on the left of group A, denoted as R_A, whereas, group A is exerting a force to the right on the rope, denoted as A_R. Likewise, group A is exerting force to the left on the earth, whereas, the earth is exerting force to the right on the group A. These two examples are of action and reaction forces.

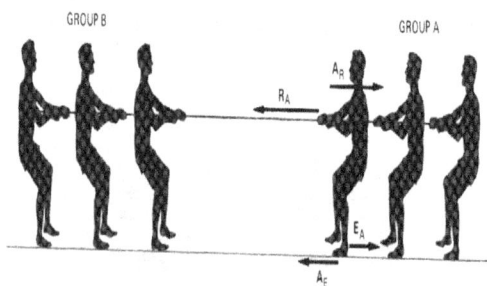

Figure 2.2. In tug of war competition, the forces being exerted on the team members by rope as well as vice versa are examples of Newton's third law of motion. Besides, the force which the group members exert on the ground and vice versa is also examples of the third law of motion.

Source: http://physics.doane.edu/hpp/Resources/Fuller3/pdf/F3Chapter_4.pdf.

It must be noted in the above examples that each force has both magnitude and direction, and therefore, force is a vector quantity. The action-reaction forces pair is vectors having equivalent magnitude but opposite in direction. A vector representation of the action-reaction pair of forces for the tug of war competition in Figure 2.2 can be given as: $F_{AB} = - F_{BA}$. The rules of vector subtraction and addition can be applied to a force system.

In agreement with the *equilibrium* definition, an object at rest does not experience any net force.

The vector addition of all forces acting upon an object in mechanical equilibrium is zero.

Another approach of describing equilibrium is that the vector addition of the components of a force in any direction is equivalent to zero. Let us consider a particular scenario. The leg in Figure 2.3 is held in power by the cord *OA* and the weight denoted by *W*. Reflect on the forces acting upon the foot at point *O*. We will denote the tension in the cord *OA* as T_A; the tension in the leg as T_B, whereas, the tension in the cord *OC* is denoted as T_C. The tension T_C is equivalent to the weight, which says 10.0 Newtons (N). The foremost step in finding the solution to any mechanical equilibrium problem is drawing a diagram of force like the one shown below. A force diagram is a naivety of the physical problem. We draw all the forces acting upon the system. A vector represents each force. The length of the arrow corresponds to the magnitude of the force; whereupon, the direction of the arrow denotes the direction in which the force is acting. In numerous cases, drawing an accurate force diagram will principally solve a force problem for us. The best means in order to learn drawing of force diagrams is to practice it (Keller, 1942; Niaz, 1995).

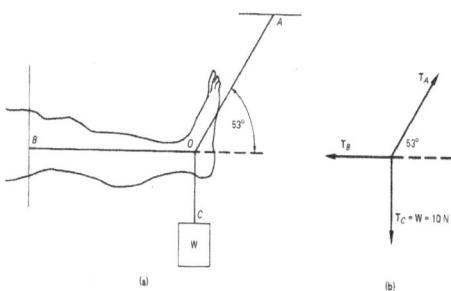

Figure 2.3. Schematic representation of a pulley-traction device. (a) The leg is in state of equilibrium; (b) The resulting force is zero at point O.

The addition of the components in the vertical direction:

Vertical components of force = $T_A \sin 53° - W = 0$. Substituting the values for the sine and cosine of 53°, and $T_A = T_B$, and $0.800\ T_A = 10.0$.

By solving the above two equations, we get the values for the magnitudes of both T_A and T_B:

$T_A = 12.5\ N$

whereas,

$T_B = 7.50\ N$

The directions of both forces are shown in Figure 2.3. T_B is acting horizontally to the left; whereas, T_A is acting to the right in the upward direction (53° above horizontal).

2.3. NEWTON'S SECOND LAW OF MOTION

Consider the case of moving furniture. The huge pieces of furniture involve a great amount of force to be exerted on them in order to get them to move. The smaller furniture pieces can be moved more quickly as we increase the force exerted upon them (Buchdahl, 1951; Milton and Willis, 2007). These experiences might make us willing to accept the assertion that the force needed to be exerted on an object to gain certain acceleration is proportional to the mass of the object; moreover, the amount of acceleration which we succeed in providing to an object is proportional to the force which is exerted upon that object (ignoring the effects of friction). There is unbalanced force acting upon the objects which we move. Resultantly, the objects are no more in a state of rest and are not moving at constant velocity. On the other hand, they are traveling with a velocity (Romero et al., 2003; Gundlach et al., 2007). As explained by Newton's second law of motion, such objects have acceleration in the direction of the unbalanced force.

The time rate of change in velocity of any object is proportionate to the total unbalanced force acting on it.

We can explain our data by assuming that the force needed to produce a given amount of acceleration is proportionate to the mass (Plastino and Muzzio, 1992; Wayne, 2012):

$F \propto m$ for constant acceleration

For a specified mass of a body, the acceleration is proportionate to the amount of force $a \propto F$.

We can combine these two proportionalities into one equation of the form $F = kma$. In this equation, k is proportionality constant. We can make k unity if we select two of the variable units and then define the third in terms of these two variables. That is the usual approach.

In SI, the Newton is defined as the magnitude of force which provides a mass of 1 kg an acceleration of 1 meter per second squared (m/sec²). Hence, the vector equation for Newton's second law is stated as:

$\mathbf{F} = m\mathbf{a}$; therefore, $\mathbf{F}$ and $\mathbf{a}$ are parallel   (1)

Here, $\mathbf{F}$ is the net force measured in Newtons, m is the mass in kg; whereas, $\mathbf{a}$ is the acceleration in meters per second squared. Then the weight of a body in N can be calculated by multiplication of the mass of the body in kg with the acceleration due to gravity, denoted as g, and having unit meters per second squared. $\mathbf{W} = m\mathbf{g}$. The mass of an object remains the same at every place. Since the acceleration due to gravity is dependent on mass position on the earth, therefore, the weight of the object can vary accordingly. For the numerical calculations, we employ a value of 9.80 m/sec² for g close the surface of the earth. According to Newton's 2nd law of motion, the effect of a force is to change the state of motion of a body (Arminjon, 2006; Pourciau, 2006; Ignatiev, 2007) (Figure 2.4).

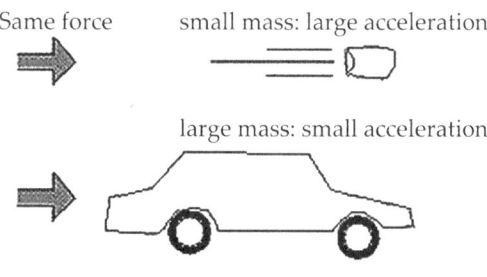

Same force   small mass: large acceleration

large mass: small acceleration

Force = mass x acceleration

Figure 2.4. Graphic representation of Newton's second law.

Source: https://www.pinterest.com/pin/308778118171352303/?lp = true.

2.4. FORCE OF FRICTION

In all cases of real motion, there exists a force that opposes the motion. This force is known as the force of friction. The properties of this frictional force might be difficult to understand; moreover, the source of the frictional force might be vague (Marti et al., 1990; Bhushan and Sundararajan, 1998). Consider an experiment of sliding one surface over another surface. A few of the variables we might alter are the various conditions of surfaces, types of surfaces, the velocity, and the area of contact. After that, we compare the initial force and the force for uniform motion and tabularize our findings (Colchero et al., 1992; Fujihira et al., 1996).

We might note the following observations during this experiment:

- The initial force is greater than the force needed to continue an object moving in a uniform manner after it is started.
- The frictional force is not dependant on the velocity (in case of low velocities only).
- The frictional force is not dependant on the area of contact.
- The frictional force is reliant on the forces pressing the surfaces together.
- The frictional force depends upon the types of surfaces that are in contact.
- The frictional force depends on the condition of the surfaces that are in contact.

In order to explain such observations, we suppose that we can represent the force of friction as a force having magnitude equivalent to a constant number multiplied by the force which is pushing the two surfaces together. Mathematically, we can write it as:

$$f_s \leq \mu_s N \qquad (2)$$

In this case, f_s is the *static frictional force*, whereas μ_s is known as the *coefficient of static friction*, and N is the normal force acting perpendicular to the surfaces and pressing the surfaces together. The direction of the force of friction is opposite to the direction of imminent motion. The static friction force can have any value from 0 to $\mu_s N$, its maximum value is just prior to the beginning of the motion (Labardi et al., 1994; Su et al., 2013).

The coefficient of friction is maximum among two surfaces which are in a state of rest with respect to one another. As the two surfaces starts to move with respect to each other, the force of friction starts decreasing. The moment when sliding starts, a *kinetic frictional force* starts opposing this sliding motion. For this force, the coefficient of friction is the *coefficient of kinetic friction* μ_k. We write the kinetic frictional force as the product of the normal force, N, pressing the surfaces together and μ_k (Figure 2.5).

$$f_k = \mu_k N \qquad\qquad (3)$$

Figure 2.5. Real-life example of friction force.

Source: https://k8schoollessons.com/friction-for-kids/.

Table 2.1 provides some approximate values of μ_s and μ_k. It must be noted that for each case $\mu_k < \mu_s$, that is, the static friction is more than the kinetic friction.

In numerous cases, we desire to reduce the frictional force to a minimum. In different mechanical systems, the contact surfaces are polished in order to decrease the frictional forces. One more method of reducing the effects of friction is by lubricating the surfaces. In the case of the joints and membranes in the human body, the problem is also resolved by using lubricants (Schitter and Stemmer, 2003; Li et al., 2006).

Table 2.1. Few Expected Values of μs and μk

Surfaces	μ_k	μ_s
Automobile tire on clean concrete	0.8	1.0
Automobile tire on muddy concrete	0.2	0.7

Automobile tire on icy concrete	0.02	0.3
Wood on wood	0.3	0.5
Leather on metal	0.5	0.6
Metal on metal (dry)	0.2	0.6
Metal on metal (lubricated)	0.04	0.1
Metals on wood	0.2	0.6
Joints of human body	0.0016 to 0.005	0.02

In Table 2.1, note the value of μ in the human body. It raises a question of how the joints in the human body can have such a slight value of coefficient of friction. The synovial joint provides the answer to this question. The ends of the bones in a synovial joint are covered by cartilage and rest against an enclosed membrane sack. The synovial fluid is present inside the membrane sack which in its properties is similar to the blood plasma. This synovial fluid aids in supporting the forces exerted by the bones and acts as a medium to decrease the frictional forces (El-Shimi, 1977; Pettersson et al., 2007).

2.5. THE LAW OF ATTRACTION AMONG TWO BODIES

A well-known fact is that a freely falling body has an acceleration g. In harmony with Newton's 2nd law of motion, there must be an unbalanced force which acts on the falling body. This particular force is the weight of that body and is a consequence of the attraction between the earth and that falling body (Hamilton, 1847; Reynolds, 1885). The law of gravitational attraction among two idealized point masses as described by Sir Isaac Newton states that:

"Any two point masses attract each other with a force proportionate to the product of their masses and inversely proportionate to the square of the distance among the masses. The direction of such force is along the line among the point masses."

This law is known as the *universal law of gravitation* as it is meant to apply to point mass interactions all over the universe. Mathematically, we can write this law as:

$$F = G\,m_1 m_2 / r^2 \qquad (4)$$

Here, G is a constant recognized as the *universal gravitation constant*. The value of G is the same for any pair of masses. In SI units, the value of G

= 6.6732 x 10^{-11} N- m^2/kg^2. This law can be extended to all spherical bodies having mass distributions depending just on distance from the center of the body (Tomlinson, 1929; Boersma, 1961; Yu and Polycarpou, 2004).

Suppose a mass, m, is falling freely near the surface of the earth. The force of gravity gives rise to the acceleration g. Therefore, we have: (r_e = earth radius *and* M_e = mass of the Earth)

$$mg = GM_e m/r_e^2 \text{ or } g = GM_e/r_e^2 \tag{5}$$

Gravitation is yet one more example of interaction-at-a-distance as an object needs not to be in direct contact with the earth in order to be acted upon by the force of gravity. The concept of a gravitational field surrounding the earth is quite reasonable. We can define the magnitude of the gravitational field, at a distance r from the center of the earth, as GM_e/r^2 Newtons, and having a direction toward the center of the earth. Placing an object having mass m at a distance r from the center of the Earth, we will have its weight W, or the gravitational force acting upon it, equal to GmM_e/r^2 Newtons (Derjaguin et al., 1956; Sokolskii and Sadovnikov, 1987). The weight of the body changes as it moves away from the earth. This is due to a change in the strength of the gravitational field with the change in the distance from the center of earth. Moreover, the gravitational fields due to other objects, for instance, the sun, and moon, can be used to explain different phenomena such as planetary orbits and tides (Lifshitz and Hamermesh, 1992; Dorofeyev, 1998; Kantorowicz, 2016).

2.6. CENTRIPETAL FORCE

In case of uniform circular motion, there exists an acceleration that is directed towards the center. Newton's 2nd law of motion states that an acceleration produced in a body is a result of an unbalanced force acting upon it. The vectors of the unbalanced force and acceleration are directed toward the center of the circle (Hubbard, 1995, 1996). In uniform circular motion, we can write the magnitude of the unbalanced force acting upon the body towards the center of rotation as below:

$$F = m\, a_r = mv^2/r \tag{6}$$

The speed of an object experiencing a circular motion is equivalent to $2\pi rn$, where r is the radius of the circle, and n is the number of revolutions per sec. we can substitute this expression for v in the above equation to get the following result,

$$F = 4\pi^2\,n^2\,mr \tag{7}$$

This force is known as the centripetal force. In the above equation, the expression $4\pi^2\,n^2$ represents the square of $2\pi n$, where 2π is the total radians in one revolution (2π radians $= 360°$). The quantity $2\pi n$ is known as the *angular velocity*. Angular velocity is measured in radians per second and is denoted by ω.

$$\omega = 2\pi n$$

So, we can rewrite Eq. (7) as:

$$F = mr\omega^2 \tag{8}$$

In this age of space, we are quite aware of the idea of weightlessness as has been experienced by astronauts in orbiting satellites. The gravitational attraction of the earth for these people as well as for the entire contents of the spacecraft is exactly equivalent to the centripetal force of the body in its orbit. By putting values, we get the following equation:

$$F = G\,M_e\,m/r^2 = m\,v^2/r \tag{9}$$

In this equation, G is the gravitation constant, M_e is the mass of the earth, whereas, r is the radius of the orbit. Therefore, the speed of the spacecraft in orbit is assumed by (Figure 2.6):

$$v = \sqrt{\frac{GM_e}{r}} \tag{10}$$

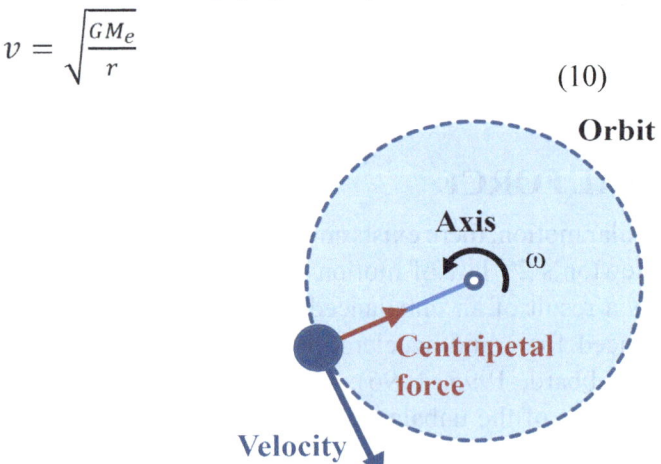

Figure 2.6. Representation of centripetal force.

Source: https://en.wikipedia.org/wiki/Centripetal_force.

There is a difference in experience between the orbiting astronauts and the observers on the ground (Vartholomeos and Papadopoulos, 2006; Jing and Jian-Chao, 2010). What we experience on earth is that the astronauts, as well as all the orbiting equipment, are falling around the earth and are being acted upon by a centripetal force whose magnitude is determined by Equation (8). On the other hand, in the orbiting spacecraft, all objects and equipment are at rest with respect to each other. While stepping away from the dinner table, the astronaut does not fall toward the earth however continue in an orbit around the earth. For an astronaut, the force of gravity on objects gets canceled by an fictitious force, known as *centrifugal force*, which arises owing to the orbital motion of spacecraft (Graybiel and Johnson, 1963; Erlichson, 1991; Kerzel, 2003). In reality, the astronauts are just a few hundred km above the earth's surface. The force of gravity at that height is decreased by only around 10%. Therefore, the weightlessness experience takes place in the frame of reference of those astronauts who can justify the floating objects in the spacecraft by supposing that these objects are being acted upon by an imaginary outward force equivalent in magnitude to the force of gravity given by Equation (8).

2.7. GRAVITATIONAL AND INERTIAL MASS EQUIVALENCE

Newton raised a basic question regarding the inertia of an object. He deliberated if the property responsible for the weight of a body was similar to the inertial parameter determining the acceleration formed due to a force applied on the object (Nordtvedt Jr, 1968; Braginskii and Panov, 1971; Koester, 1976). Another way of stating this is: Is the inertial mass of an object the same as the gravitational mass of that object? In the case of freely falling objects, does the acceleration depends on the properties of the falling object or does it remain constant? In the case of the inertial mass being equivalent to the gravitational mass, we should suppose all objects to fall with the equivalent acceleration. Similarly, if a nongravitational force equivalent to the weight of the object is exerted, the acceleration 'a' of the object must be equivalent to the acceleration because of gravity (Roll et al., 1964; Ohanian, 1971, 1977; Hynecek, 2005). Since the 17th century, many such experiments have been carried out, and all such experiments exhibit that within experimental error, the inertial mass and the gravitational mass have the same value (Adunas et al., 2001; Hui and Nicolis, 2010; Kajari et al., 2010).

2.8. ESCAPE VELOCITY FROM THE EARTH

Considering the mass as constant, the force is the product of acceleration and mass. Generally, an equation of motion is stated as (Ahrens and O'Keefe, 1987; Ahrens and Harris, 1992; Tian et al., 2005)

$$F_{net} = ma = m \, dv/dt = m \, d^2r/dt^2 \tag{11}$$

In this equation, the acceleration a is specified by the time rate of change of the velocity (in this case the speed, since we are dealing just with the modulus of the position), that is $a = dv/dt$, whereas, the speed v is specified by the time rate of change of the distance, written as, $v = dr/dt$.

In a given scenario, provided we know the expression for F, we can equate it to $m \, d^2r/dt^2$, and get the solution of the differential equation (Frank, 1947; Wetherill, 1976). There might be several functions of F of time t or distance r. For instance, F might be stated as the law of gravitational attraction of the earth,

$$GM_e m/r^2 = m \, d^2r/dt^2 \tag{12}$$

In order to solve this differential equation, we can take a mnemotechnic approach, we express the acceleration as a function of the distance, derivative of the speed times dr/dr and then reordering the variables:

$$a = dv/dt = dv/dt \cdot dr/dr = dr/dt \cdot dv/dr = v \, (dv/dr) \tag{13}$$

We may write Newton's second law for constant mass as:

$$F = m \, v \, dv/dr \tag{14}$$

The above expression can be set equivalent to the gravitational attraction of the earth,

$$GM_e m/r^2 = m \, v \, dv/dr \tag{15}$$

In this equation, M_e is the mass of the earth. We can get an expression for the least vertical velocity required for an object in order to escape from the earth by integrating this equation (Hendel, 1983; Canup, 2008).

The constant for the escape velocity can be assessed by equating the velocity of the object to be 0 when it is at an immeasurable distance from the earth. The constant then will be zero. Therefore, the vertical velocity required by an object at the surface of the earth in order to escape from the earth is given by:

$$v_{escape} = \sqrt{\frac{2GM_e}{r_e}} \tag{16}$$

Here, r_e denotes the radius of the earth.

REFERENCES

1. Adunas, G. Z., Rodriguez-Milla, E., & Ahluwalia, D. V., (2001). Probing quantum violations of the equivalence principle. *General Relativity and Gravitation, 33*(2), 183–194.

2. Ahrens, T. J., & Harris, A. W., (1992). Deflection and fragmentation of near-Earth asteroids. *Nature, 360*(6403), 429.

3. Ahrens, T. J., & O'Keefe, J. D., (1987). Impact on the earth, ocean, and atmosphere. *International Journal of Impact Engineering, 5*(1–4), 13–32.

4. Anderson, J. L., (1990). Newton's first two laws of motion are not definitions. *American Journal of Physics, 58*(12), 1192–1195.

5. Arminjon, M., (2006). On the extension of Newton's second law to theories of gravitation in curved space-time. *Arch. Mech., 48*(gr-qc/0609051), 551–576.

6. Bhushan, B., & Sundararajan, S., (1998). Micro/nanoscale friction and wear mechanisms of thin films using atomic force and friction force microscopy. *Acta Materialia, 46*(11), 3793–3804.

7. Boersma, J., (1961). Mathematical theory of the two-body problem with one of the masses decreasing with time. *Bulletin of the Astronomical Institutes of the Netherlands, 15*, 291–301.

8. Braginskii, V. B., & Panov, V. I., (1971). Verification of the equivalence of the inertial and the gravitational mass (Inertial and gravitational mass equivalence principle verified for aluminum and platinum, using torsional pendulum with large relaxation time). *Zhurnal Eksperimental'noi i Teoreticheskoi Fiziki, 61*, 873–879.

9. Brown, D. E., (1989). Students' concept of force: The importance of understanding Newton's third law. *Physics Education, 24*(6), 353.

10. Buchdahl, G., (1951). Science and logic: Some thoughts on Newton's second law of motion in classical mechanics. *The British Journal for the Philosophy of Science, 2*(7), 217–235.

11. Canup, R. M., (2008). Accretion of the earth. *Philosophical Transactions of the Royal Society A: Mathematical, Physical and Engineering Sciences, 366*(1883), 4061–4075.

12. Cohen, I. B., (1987). Newton's third law and universal gravity. *Journal of the History of Ideas, 48*(4), 571–593.

13. Colchero, J., Bielefeldt, H., Ruf, A., Hipp, M., Marti, O., & Mlynek,

J., (1992). Scanning force and friction microscopy. *Physica. Status Solidi(a)*, *131*(1), 73–75.

14. Derjaguin, B. V., Abrikosova, I. I., & Lifshitz, E. M., (1956). Direct measurement of molecular attraction between solids separated by a narrow gap. *Quarterly Reviews, Chemical Society*, *10*(3), 295–329.

15. Dorofeyev, I. A., (1998). The force of attraction between two solids with different temperatures. *Journal of Physics A: Mathematical and General*, *31*(19), 4369.

16. El-Shimi, A. F., (1977). In vivo skin friction measurements. *J. Soc. Cosmet. Chem.*, *28*(2), 37–52.

17. Erlichson, H., (1991). Motive force and centripetal force in Newton's mechanics. *American Journal of Physics*, *59*(9), 842–849.

18. Felten, J. E., (1984). Milgrom's revision of Newton's laws-dynamical and cosmological consequences. *The Astrophysical Journal*, *286*, 3–6.

19. Frank, J. M., (1947). The problem of escape from the earth by rocket. *Journal of the Aeronautical Sciences*, *14*(8), 471–480.

20. Fujihira, M., Aoki, D., Okabe, Y., Takano, H., Hokari, H., Frommer, J., & Sakai, F., (1996). Effect of capillary force on friction force microscopy: A scanning hydrophilicity microscope. *Chemistry Letters*, *25*(7), 499–500.

21. Galili, I., & Tseitlin, M., (2003). Newton's first law: Text, translations, interpretations, and physics education. *Science and Education*, *12*(1), 45–73.

22. Graybiel, A., & Johnson, W. H., (1963). XXVII A comparison of the symptomatology experienced by healthy persons and subjects with loss of labyrinthine function when exposed to unusual patterns of centripetal force in a counter-rotating room. *Annals of Otology, Rhinology and Laryngology*, *72*(2), 357–373.

23. Gundlach, J. H., Schlamminger, S., Spitzer, C. D., Choi, K. Y., Woodahl, B. A., Coy, J. J., & Fischbach, E., (2007). Laboratory test of Newton's second law for small accelerations. *Physical Review Letters*, *98*(15), 150801.

24. Hamilton, W. R., (1847). The hodograph or a new method of expressing in symbolical language the Newtonian law of attraction. In: *Proceedings of the Royal Irish Academy* (Vol. 3, pp. 344–353).

25. Handhika, J., Cari, C., Soeparmi, A., & Sunarno, W., (2016). Student conception and perception of Newton's law. In: *AIP Conference Proceedings* (Vol. 1708, No. 1, p. 070005). AIP Publishing.

26. Hecht, E., (2015). Origins of Newton's first law. *The Physics Teacher*, *53*(2), 80–83.

27. Hellingman, C., (1992). Newton's third law revisited. *Physics Education*, *27*(2), 112.

28. Hendel, A. Z., (1983). Solar escape. *American Journal of Physics*, *51*(8), 746–748.

29. Hubbard, T. L., (1995). Environmental invariants in the representation of motion: Implied dynamics and representational momentum, gravity, friction, and centripetal force. *Psychonomic Bulletin and Review*, *2*(3), 322–338.

30. Hubbard, T. L., (1996). Representational momentum, centripetal force, and curvilinear impetus. *Journal of Experimental Psychology: Learning, Memory, and Cognition*, *22*(4), 1049.

31. Hui, L., & Nicolis, A., (2010). Equivalence principle for scalar forces. *Physical Review Letters*, *105*(23), 231101.

32. Hynecek, J., (2005). Remarks on the equivalence of inertial and gravitational masses and on the accuracy of Einstein's theory of gravity. *Physics Essays*, *18*(2), 1–18.

33. Ignatiev, A. Y., (2007). Is violation of Newton's second law possible? *Physical Review Letters*, *98*(10), 101101.

34. Itza-Ortiz, S. F., Rebello, S., & Zollman, D., (2003). Students' models of Newton's second law in mechanics and electromagnetism. *European Journal of Physics*, *25*(1), 81.

35. Jing, Z., & Jian-Chao, Z., (2010). A virtual centripetal force-based coverage-enhancing algorithm for wireless multimedia sensor networks. *IEEE Sensors Journal*, *10*(8), 1328–1334.

36. Kajari, E., Harshman, N. L., Rasel, E. M., Stenholm, S., Süßmann, G., & Schleich, W. P., (2010). Inertial and gravitational mass in quantum mechanics. *Applied Physics B*, *100*(1), 43–60.

37. Kantorowicz, E., (2016). *The King's Two Bodies: A Study in Medieval Political Theology* (Vol. 22, pp. 1–20). Princeton University Press.

38. Keller, J. M., (1942). Newton's third law and electrodynamics.

American Journal of Physics, 10(6), 302–307.

39. Kerzel, D., (2003). Centripetal force draws the eyes, not memory of the target, toward the center. *Journal of Experimental Psychology: Learning, Memory, and Cognition, 29*(3), 458.

40. Koester, L., (1976). Verification of the equivalence of gravitational and inertial mass for the neutron. *Physical Review D, 14*(4), 907.

41. Kurniawan, Y., Suhandi, A., & Hasanah, L., (2016). The influence of implementation of interactive lecture demonstrations (ILD) conceptual change oriented toward the decreasing of the quantity students that misconception on the Newton's first law. In: *AIP Conference Proceedings* (Vol. 1708, No. 1, p. 070007). AIP Publishing.

42. Labardi, M., Allegrini, M., Salerno, M., Frediani, C., & Ascoli, C., (1994). Dynamical friction coefficient maps using a scanning force and friction microscope. *Applied Physics A, 59*(1), 3–10.

43. Li, H., Furuta, K., & Chernousko, F. L., (2006). Motion generation of the capsubot using internal force and static friction. In: *Proceedings of the 45th IEEE Conference on Decision and Control* (Vol. 45, pp. 6575–6580). IEEE.

44. Lifshitz, E. M., & Hamermesh, M., (1992). The theory of molecular attractive forces between solids. In: *Perspectives in Theoretical Physics* (Vol. 1, pp. 329–349). Pergamon.

45. Low, D. J., & Wilson, K. F., (2017). The role of competing knowledge structures in undermining learning: Newton's second and third laws. *American Journal of Physics, 85*(1), 54–65.

46. MaKinster, J. G., Beghetto, R. A., & Plucker, J. A., (2002). Why can't I find Newton's third law? Case studies of students' use of the web as a science resource. *Journal of Science Education and Technology, 11*(2), 155–172.

47. Maloney, D. P., (1984). Rule-governed approaches to physics-Newton's third law. *Physics Education, 19*(1), 37.

48. Marti, O., Colchero, J., & Mlynek, J., (1990). Combined scanning force and friction microscopy of mica. *Nanotechnology, 1*(2), 141.

49. Milton, G. W., & Willis, J. R., (2007). On modifications of Newton's second law and linear continuum elastodynamics. *Proceedings of the Royal Society A: Mathematical, Physical and Engineering Sciences, 463*(2079), 855–880.

50. Minstrell, J., (1984). Teaching for the development of understanding of ideas: Forces on moving objects. *Ob-Serving Science Classrooms: Observing Science Perspectives from Research and Practice, 1*(1), 55–74.

51. Niaz, M., (1995). Chemical equilibrium and Newton's third law of motion: Ontogeny/phylogeny revisited. *Interchange, 26*(1), 19–32.

52. Norbury, J. W., (1998). From Newton's laws to the wheeler-DeWitt equation. *European Journal of Physics, 19*(2), 143.

53. Nordtvedt, Jr. K., (1968). Equivalence principle for massive bodies. I. phenomenology. *Physical Review, 169*(5), 1014.

54. Ogborn, J., & Taylor, E. F., (2005). Quantum physics explains Newton's laws of motion. *Physics Education, 40*(1), 26.

55. Ohanian, H. C., (1971). Inertial and gravitational mass in the Brans-Dicke theory. *Annals of Physics, 67*(2), 648–661.

56. Ohanian, H. C., (1977). What is the principle of equivalence? *American Journal of Physics, 45*(10), 903–909.

57. Pettersson, T., Nordgren, N., Rutland, M. W., & Feiler, A., (2007). Comparison of different methods to calibrate torsional spring constant and photodetector for atomic force microscopy friction measurements in air and liquid. *Review of Scientific Instruments, 78*(9), 093702.

58. Pfister, H., (2004). Newton's first law revisited. *Foundations of Physics Letters, 17*(1), 49–64.

59. Plastino, A. R., & Muzzio, J. C., (1992). On the use and abuse of Newton's second law for variable mass problems. *Celestial Mechanics and Dynamical Astronomy, 53*(3), 227–232.

60. Pourciau, B., (2006). Newton's interpretation of Newton's second law. *Archive for History of Exact Sciences, 60*(2), 157–207.

61. Pugh, K. J., (2004). Newton's laws beyond the classroom walls. *Science Education, 88*(2), 182–196.

62. Reynolds, O., (1885). LVII. On the dilatancy of media composed of rigid particles in contact. With experimental illustrations. *The London, Edinburgh, and Dublin Philosophical Magazine and Journal of Science, 20*(127), 469–481.

63. Roberson, W. C., Gallagher, J., & Miller, W., (2004). Newton's first law: Not so simple after all. *Science and Children, 41*(6), 25.

64. Roll, P. G., Krotkov, R., & Dicke, R. H., (1964). The equivalence of inertial and passive gravitational mass. *Annals of Physics*, *26*(3), 442–517.

65. Romero, J. M., Santiago, J. A., & Vergara, J. D., (2003). Newton's second law in a non-commutative space. *Physics Letters A*, *310*(1), 9–12.

66. Santavy, I., (1986). Newton's first law. *European Journal of Physics*, *7*(2), 132.

67. Savinainen, A., Scott, P., & Viiri, J., (2005). Using a bridging representation and social interactions to foster conceptual change: Designing and evaluating an instructional sequence for Newton's third law. *Science Education*, *89*(2), 175–195.

68. Schitter, G., & Stemmer, A., (2003). Model-based signal conditioning for high-speed atomic force and friction force microscopy. *Microelectronic Engineering*, *67*, 938–944.

69. Sokolskii, A. G., & Sadovnikov, A. A., (1987). Lagrangian solutions for weber's law of attraction. *Soviet Astronomy*, *31*, 90.

70. Su, H., Wu, C. S., Pittner, A., & Rethmeier, M., (2013). Simultaneous measurement of tool torque, traverse force and axial force in friction stir welding. *Journal of Manufacturing Processes*, *15*(4), 495–500.

71. Terry, C., & Jones, G., (1986). Alternative frameworks: Newton's third law and conceptual change. *European Journal of Science Education*, *8*(3), 291–298.

72. Thornton, R. K., & Sokoloff, D. R., (1998). Assessing student learning of Newton's laws: The force and motion conceptual evaluation and the evaluation of active learning laboratory and lecture curricula. *American Journal of Physics*, *66*(4), 338–352.

73. Tian, F., Toon, O. B., Pavlov, A. A., & De Sterck, H., (2005). A hydrogen-rich early Earth atmosphere. *Science*, *308*(5724), 1014–1017.

74. Tinto, V., (2013). Isaac Newton and student college completion. *Journal of College Student Retention: Research, Theory and Practice*, *15*(1), 1–7.

75. Tomlinson, G. A., (1929). CVI. A molecular theory of friction. *The London, Edinburgh, and Dublin Philosophical Magazine and Journal of Science*, *7*(46), 905–939.

76. Vartholomeos, P., & Papadopoulos, E., (2006). Analysis, design and control of a planar micro-robot driven by two centripetal-force

actuators. In: *Proceedings 2006 IEEE International Conference on Robotics and Automation* (Vol. 1, pp. 649–654). ICRA, IEEE.

77. Wayne, R., (2012). A fundamental, relativistic and irreversible law of motion: A unification of Newton's second law of motion and the second law of thermodynamics. *The African Review of Physics, 7(1), 1–19.*

78. Wetherill, G. W., (1976). The role of large bodies in the formation of the Earth and Moon. In: *Lunar and Planetary Science Conference Proceedings* (Vol. 7, pp. 3245–3257).

79. White, B. Y., (1984). Designing computer games to help physics students understand Newton's laws of motion. *Cognition and instruction, 1*(1), 69–108.

80. Yu, N., & Polycarpou, A. A., (2004). Adhesive contact based on the Lennard-Jones potential: A correction to the value of the equilibrium distance as used in the potential. *Journal of Colloid and Interface Science, 278*(2), 428–435.

Chapter 3

Space, Time, and Motion

CONTENTS

3.1. INTRODUCTION

Physics is a discipline that aims to provide a quantitative, mathematical explanation of natural phenomena. This means that the physical laws are generally mathematical relations amongst the physical quantities. These relations are considered correct only if they relate to experience. Physical laws should always be verified experimentally (Galton, 1993; Banuelos and Méndez-Hernández, 2003). An experiment is the only arbiter of the scientific truth. Subsequently, any physical quantity should be measurable, namely a particular set of operations that need to be executed to measure it should be defined. First, the system of the units of measurement should be defined. The first three sections outline how this is done. The selection of units is the priority arbitrary; physical laws are dependent of the nature of the phenomena, not on our selections (Harichandran and Vanmarcke, 1986; Rose et al., 1996; Laptev, 2005). In practice, conversely, having standardized selections is especially important to make the outcomes understandable to everybody. International agreements have a well-defined system of units to be named, Système International, in French (Muller, 1998).

Few physical quantities, like temperature and mass, are represented by a single number and are known as a scalar. Other, like force and velocity, are more complex; identifying how big they are isn't enough, also their direction needs to be given (Rude, 1997; Gorelick et al., 2007; Shechtman and Irani, 2007). Mathematically, a well-ordered set of the real numbers signifies them; they are known as vector quantities. In the following sections, some elements will be introduced that will be helpful in the following on one more mathematical object, matrices (Toulmin, 1959; Van der Vegt and Van der Ven, 2002; Nespor, 2014).

In the second part of this chapter physics, dealing with kinematics of the point-like particle, the study of the motion of particle, independently of its reasons will be described. The concepts of acceleration, velocity, and angular velocity will be outlined. These are all vector quantities, generally depending on time (Alexander, 1920; Rynasiewicz, 1995; Cremers and Soatto, 2003).

3.2. MEASUREMENT OF THE PHYSICAL QUANTI-TIES

Physics provides a quantitative description of natural phenomena. Measurement of appropriate physical quantities leads to the discovery of physical laws, which are the mathematical relations between those quantities (Whitney, 1968; Niu et al., 1976; Valassi, 2003).

All-natural phenomena occur in space and have the temporal duration; few of them occur before, others afterward. Consequently, time and space are fundamental concepts. Physical objects are usually characterized by quantities like area, length, volume, mass, color, hardness, temperature, etc. All these concepts result from the common experience and are proper of the common language (Einstein et al., 1935; Busch, 1991).

However, Physics should give a rigorous definition to every quantity, to be able to provide its numerical values. In this process of definition, the concept might become moderately different from the common language (Mäntylä and Koponen, 2007; Woodard and Taylor, 2007).

For instance, consider the length of the object or distance between the two places. If one wants to designate the number one must first define the unit of length. Indeed, one will say: "That bar is five-meter long," or, if one is in England: "That city is twenty miles away." The measure of length of the object is ratio amongst its length and length of another object one has chosen as the unit. "A bar is five-meter long," means that the length of bar is equal to the five one-meter long rules in the line (Cassinelli and Lahti, 1989; Liehr et al., 2017).

The measurement of the physical quantity is a ratio amongst that particular quantity and its measurement unit.

The measurement operation permits relating a number to every physical quantity. The symbols that look in the physical laws signifying the several physical quantities are just the numbers. For instance, when it is written $F = ma$ it means that the ratio amongst the force considered and the force is taken as a unit, corresponds to the ratio of mass of the object and mass is taken as the unit, times the ratio amongst the designated and unit acceleration (Gamba, 1967; Zhang et al., 1995).

Every physical quantity should be measurable, and the definition must be accurate and rigorous. The most effective method to define the physical quantity is the operational definition. This is described *as a set of the operations required to measure that particular quantity.*

This process has two significant implications. The first association is that the quantities that aren't measurable aren't physical quantities. This doesn't infer that such quantities can't be used. Indeed, they are frequently helpful in mathematical developments of the theory. Any theory, conversely, if it's the physical theory, should lead to estimates that are testable experimentally. The experimentally testable calculations are mathematical relations between the physical quantities, meaning the measurable ones. The secondary, non-measurable, quantities shouldn't appear in the final theoretical expression (Beltrametti et al., 1990; Poghossian et al., 2003).

The second association is that the scale matters: quantities might be large or small. Consider for instance the length. If one wants to measure the distances, say, from mm (millimeters) to km (kilometers), one can utilize graduated bars or the rules. If one needs to measure the distances of 10s or 100s of km, as for instance amongst the two mountaintops or height of Mount Everest, the process is different and one must perform triangulations. If distances are quite larger, as the distances of galaxies, the processes change completely again. And different processes are needed to measure the small distances like diameter of an atomic nucleus or of an atom. In every range of the orders of magnitude, set of procedures to measure the length is different. To be severe one would require talking of numerous different lengths. This would cause terrible confusion. Fortunately, it is experimentally verified that, in large intervals where two or more approaches work contemporarily, the outcomes are equal, and one can define the single length concept. However, the arguments above tell to be extra careful. Suppose that the physical law is very well verified experimentally for objects of the sizes between meters and km. One inclines to ponder the same law to be also valid for objects much larger and much smaller than that. But there isn't any guarantee that the judgment is true. On that, only the experiment can validate. For instance, the Newton laws valid at speeds of the ordinary experience aren't valid at speeds equivalent to speed of the light. The laws of classical mechanics are not valid for smaller objects and atoms (Rossow and Zhang, 1995; Döring and Isham, 2008).

The measurement unit is required for each quantity. The selection is in principle random but is very far from being used in practice. If every country, for instance, would select the different units for areas or lengths, the exchanges, not just the scientific ones, but the commercial ones too would be very complex. The units should be standardized. The problem is so significant that both procedures and units are made obligatory by law in majority of the countries.

3.3. THE INTERNATIONAL SYSTEM (SI)

The international standardization of the units triggered by the French Revolution. The Decimal Metric System was announced officially in 1791, but it took an entire century for its considerable diffusion and acceptance. In May 1875, at "Meter convention," representatives of the 17 Nations in Paris signed an international treaty. National and international laboratories were built with the task to develop measurement procedures and standards. This is an important area of physics called *metrology* (Page and Vigoureux, 1977; International Bureau of Weights and Measures et al., 2001).

International Organizations were made to substitute international standardization of the measures and weights in the world. The International Conference of Weights and Measures, CGPM for instructions utilizing the identifies in French, which happens every several years, is the key decision-making body. It decides on the advancement of the internationally accepted system of the units, named *Sistème International* in French or SI for brief. The European Community in 1971 issued an instruction to the member states for legal acceptance of the SI (Van loo et al., 1982; Mendiratta and Dimiduk, 1991).

There are two kinds of units: *derived units* and *base units*. The base units are specified by definition. Every derived unit is acquired using the laws of physics, namely the mathematical expression that associates it to the quantities of basic units. The selection of basic units and their number is, from the logic viewpoint, arbitrary. The selections are based on suitability, accepting quantities for which the measurements can be precise and reproducible (Dupree et al., 1987; Taylor and Mohr, 2001).

Let us take an example. Take physical laws:

*   Area "S" of the rectangle of sides of the lengths a & b is proportional to the product of the lengths;
*   Area "A" of the circle is proportional to the square of the length of its radius;
*   Space "s" covered by the body moving in lack of force is proportional to time t engaged and to its velocity υ. The mathematical expression is given as:

$$S = kab \quad A = k'R^2 \quad s = k''vt. \tag{1}$$

where k'', k' and k are the numerical constants. They are dependent on the selection of measurement units. One might take both area and length as the base quantities and units meter and the square foot respectively. The k' and *k* then would have definite values. The measuring system would be quite simpler taking the length as base unit, say one meter, and area as the derived unit. Still, though, some randomness remains. For instance, one can select the units in a way to have the value of $k = 1$ or, otherwise, to have the $k' = 1$. In the first choice, the unitary area is square of one-meter side, in second option it is generally the circle of one-meter radius. The second option provides $k = 1/\pi$ and seems funny. The choice $k = 1$ seems to be the apparent one, and is the one used universally, but, in principle, it isn't necessary.

Likewise, in the third equation, $k'' = 1$ by selecting as measuring unit of the velocity, velocity of the body that covers unit length in unit of the time.

As mentioned above, the internationally adopted system of the units is SI. It is the simplest to use and is the most rational one. In SI, the base units are 7:

- Length;
- Mass;
- Time;
- Electric current intensity;
- Thermodynamic temperature;
- Amount of substance;
- Luminous intensity.

For every one of them, the name of the unit and its symbol are fixed, as given in Table 3.1. Observe that the initial name of the unit is always the lower case, comprising when it's the name of the scientist. Most important, SI provides a precise and pure definition for each unit. Observe that these might change with time, as an outcome of the advancement of metrology, after official endorsement by the CGPM. The definitions of first three SI units are described below, which are the ones required in this textbook.

Table 3.1. The Seven Base Quantities Along with Their Units and Symbols

Quantity	Unit	Symbol
Length	meter/meter	M
Time	second	S
Mass	kilogram	kg
Thermodynamic temperature	Kelvin	K
Current intensity	ampere	A

Luminous intensity	candela	cd
Amount of substance	mole	mol

- Meter is the distance traveled by the light in vacuum in the time interval of 1/229792458 of a second. Kilogram (kg) is the mass of the international prototype kilogram.

- A second (s) is the duration of the 9,192,631,770 periods of radiation corresponding to transition amongst the two hyperfine levels of ground state of Cesium 133 atom.

The SI describes the names and symbols of almost all the derived units. The SI further describes the names and symbols of the multiples and the submultiples of units. This is achieved in steps generally of three orders of the magnitudes, of one order for first three, as given in Table 3.2. With exception of the k, da, and h, all of the multiple prefix symbols are the upper case; all sub-multiple prefix symbols are the lower case letters.

Table 3.2. Decimal Multiples and Sub-Multiples

Factor	Prefix	Symbol	Factor	Prefix	Symbol
10−1	deci	D	1024	yotta	Y
10−2	centi	C	1021	zetta	Z
10−3	milli	M	1018	exa	E
10−6	micro	μ	1015	peta	P
10−9	nano	N	1012	tera	T
10−12	pico	P	109	giga	G
10−15	femto	F	106	mega	M
10−18	atto	A	103	kilo	k
10−21	zepto	Z	102	hecto	h
10−24	yocto	y	10	deka	da

The derived units of measurements are defined, as already mentioned, using the physical law to have the definition as simple as feasible. Therefore, unit for the areas is square of one-meter side, unit of the volume is cube of one-meter side, unit of the velocity is velocity of the body traveling one meter in one second, etc.

Acceleration is the change of the velocity Δv divided by time interval Δt in which the change occurs, namely $a = \Delta v / \Delta t$. The acceleration unit is acceleration of the body, velocity of which changes by the unit in unit of the time. It is subsequently the m/s² or m.s⁻².

Let's now notice, as an instance, that all of the plane figures, rectangles, triangles, circles, etc. are expressed as the numerical factor time's product of the two lengths. Namely, all areas have the physical dimension of the length squared. If the unit of length is changed, for instance from meter (m) to centimeter (cm), measures of all areas vary by the same factor: 100^2 in example. The physical dimensions of the velocity are length divided by the time, of the acceleration of length divided by the time squared, etc. The conforming mathematical expressions are known as *dimensional equations* which are usually of the type:

$$[A] = [L^2], \quad [v] = [LT^{-1}], \quad [a] = [LT^{-2}] \qquad (2)$$

Dimensional equations are quite useful in practice. Consider any association amongst the physical quantities, for instance, $F = ma$ or $A + B = C$. All the terms should have same physical dimensions. Else, the variations of units will trigger the different terms to vary in different ways; the authenticity of relation would be dependent on the selection of units, which is random. This is the so-known as *homogeneity principle*. It is very helpful to check the analytical expressions acquired with less or more complex calculations (Van Assendelft et al., 1973).

It worth mentioning that there also exist physical quantities that have no dimensions, dimensionless, namely $[L\,0\,T0\,M0]$, they are just the magnitude numbers. The angle is a good example. In radians, it's the ratio amongst the arc of the circumference and its radius. If unit of the length is changed, the ratio amongst two of them doesn't change.

Finally notice that the physical law might include mathematical functions, for instance: $x = \sin \alpha, y = \exp(-\beta)$ or $z = \ln \gamma$. These expressions are valid only if the functions themselves (x, y, z) and the arguments (α, β, γ) possess no physical dimensions. All of them should be pure numbers.

3.4. SPACE AND TIME

The motion of a body indicates that its position in space changes in time. The notion of the motion is relative: a passenger in the plane sitting in the chair has fixed position comparative to plane, but moves at, say 700 km/h comparative to the person standing on the earth (Thrift, 1983; Maznevski and Chudoba, 2000). The latter moves at 700 km/h relative to the passenger, in opposite direction. A reference frame is needed to describe the motion. We usually live standing on the earth and so are the laboratories where experiments are carried out. Lets then begin by selecting the reference frame fixed on earth. Feasible choices are infinite (MacDonald and Thorne, 1982; Elith and Leathwick, 2009).

The position of the body is well-defined when it is known where it is. The simplest situation is when the particle is dealt with, a body so small that it can be well thought out point-like. It is known as the *material point*. Let's see how one can define the position of the material point. For a stretched body the orientations of all of its points must be similarly defined (Van Hove, 1954; Pickett, 1989).

To identify the position of the point in space three numbers are needed one for each of the dimensions. To describe its position on the given surface, generally, two numbers are required. To identify the position on the given curve, just one number is required (Stampfli et al., 1991; Kholodenko et al., 2010).

Let's start by taking the point P which can move only on the straight line. To locate its position:

- One of two directions have been chosen and is called positive;
- On the line, a point is chosen and is called the origin of coordinates (O in Figure 3.1a);
- A unit length is chosen. The oriented line with an origin and the measuring unit is known as the *co-ordinate axis*. The position of generic point P is specified by the real number, known as the co-ordinate of the point (x in figure), which is distance of the P from O, considered as positive if the point P is on right of the O, taken as negative if it's on the left (Eicher, 1995; Richert, 2002).

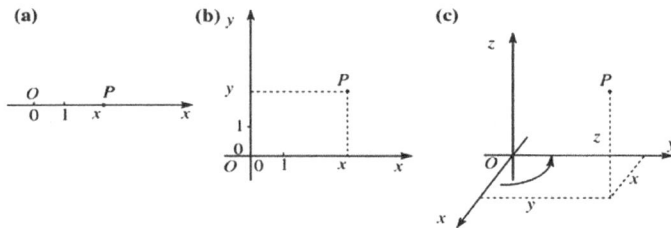

Figure 3.1. Orthogonal coordinate frames: (a) 1 dimension, (b) 2 dimensions, (c) 3 dimensions.

Source: http://www.softouch.on.ca/kb/data/Course%20in%20Classical%20 Physics%201--Mechanics%20(A).pdf.

Let's now suppose that the point P is able to move on the plane (Figure 3.1b). Two co-ordinate axes are now needed, which shouldn't be parallel. It is normally suitable to take them perpendicular, origin at a point in which the two cross and same unit length for both. The position of point P is provided by its two coordinates, which is a well-ordered pair of the real numbers (x, y).

Consider now the point in space. The reference frame displayed in Figure 3.1c is known as the *Cartesian rectangular* right-handed frame, after the René Descartes (1596–1650). It is created of three co-ordinate axes, called the x, y, and z. They cross in the single point, the origin of frame. All of the angles amongst the three pairs of the axes are right. The length units on three axes are equal. Finally, positive orientations of axes must be chosen. There are two basic possibilities. Let's assume that positive directions of the x and y are already defined. There are two probable choices for positive direction of the z. Figure 3.1c exhibits one of them; a person standing with his feet on xy plane lying along z-axis and looking down prepared to move x-axis on the y-axis by the 90° rotation, views this rotation occurring anticlockwise. The second option is opposite sign of the z. The two frames are called left-handed and right-handed respectively (Musacchio and Salmon, 2007; Jetz et al., 2012).

Now assume the axes are inverted. If it is started from the right-handed frame and one axis is inverted, that is the mirror reflection and the left-ended frame is obtained. The same occurs if all three axes are inverted. The inversion of the two axes provides, conversely, the same outcome as the rotation of 180° around third axis: initial and the final frame possess same "handedness."

To describe the reference frame a series of choices have been made, which are recalled as:

- Choice of origin;
- Choice of directions of the axes;
- Choice of positive directions;
- Choice of units.

While each of the choices is random, it can be asked whether there exists any advantaged choice, or if there exists one that is posed better, are the laws of physics independent of these specific choices? The answers can't come from mathematics or logic, but only from the experiment. Let's consider each of the choices (Snyder, 1947; Pearman et al., 2008).

- Are the laws of physics independent of the origin of the axes? In order to validate the point, let's create two identical apparatuses. Let both consist of inclined planes with balls rolling on them, flywheels, pendulums, gears, etc., all indistinguishable. The two apparatuses are positioned in two distinct locations. They are prepared to be in precisely the same initial state: pendulums are out of the equilibrium at same distance, spheres are at same heights on the planes inclined, the flywheels and gears are in the exact same positions. They are moved simultaneously and their evolutions are observed. Do the two systems evolve in same way? Do they suppose the same configurations at same time? The answer isn't always yes. However, every time difference is observed, it is probable to recognize the reason as in some physical situation that is different in two locations. For instance, the gravitational acceleration may be little different in two sites and subsequently the periods of pendulums are different too.

In any circumstance, experiments exhibit that, once all of the local effects are removed, or looked after, the two apparatuses evolve in the same manner, for example, going through same configurations at same instants. The very significant conclusion is: *"The physical laws do not depend on position."* In other words, all of the positions are the same, or space is homogeneous. Let's repeat that this is the experimental conclusion. No experiment till now has proved it wrong. It can be stated that *"physical laws are invariant,"* meaning these laws do not change, *under space translations.*

- Are laws of physics independent of the directions of axes? Let's take the two identical apparatuses and then rotate one to another. For instance, in one situation the z-axis is vertical, while in the other it is at 45° with the vertical. Do the two systems progress through the same states? Certainly not! Indeed, for instance, pendulums oscillate around the vertical axis in one situation, around an inclined axis in the other. In this situation the privileged direction exists, direction of the weight. But let us think for a moment, if we were quite far from the earth, or in the absence of weight, privileged direction wouldn't exist. That direction isn't the property of space but is "local" effect of the body, the earth. In other words, if one wants to compare the two experiments in same conditions, one should also rotate earth in the second situation. If all of the external conditions are appropriately taken into account, the experiments exhibit that *physical laws are not dependent on directions of the axes*. In other words, no advantaged direction prevails, or space is isotropic. Still in the other words, laws of *physics are invariant under the rotations*.

- Are the laws of physics, independent of orientation, right-handed or left-handed? Experiments have displayed that all laws of physics at the macroscopic level are not dependent on the choice. But this isn't true at the microscopic level. For instance, a class of the radioactive phenomena, the beta decays, known as a fundamental force or weak interaction, differentiate between the right and left handed cases. Namely, not all laws of physics are invariant under the inversion of axes.

- Are the laws of physics independent of scale of the length? This time two apparatuses are built that are indistinguishable but for having all different dimensions, scaled by some factor. Do the two evolve in a similar manner? The answer was found by Galileo Galilei and is NO.

Consider for instance beam made of certain homogeneous material. The beam has a certain length, and the cross-section, which is assumed to be rectangular, has certain width and a certain height. It is laid on the two supports close to its extremes on the horizontal plane. Suppose the beam is in equilibrium. Now take a beam similar geometrically to first one but 10 times longer, 10 times wider, and 10 times higher. Again, it is laid on two supports close to its extremes. It is observed that the beam gives out in the middle point. The reason is as the weight of beam is the force applied in the middle point directed downwards (Yoccoz et al., 2001; Hassibi and Hochwald, 2002). The weight inclines to break down the beam, the cohesion forces amongst molecules incline to keep it together. The weight, which usually is proportional to volume, is for second beam 1000 times larger as compared to the first one. The resistance to the fracture is proportional to area of cross-section and for second beam is 100 times larger as compared to the first one. Subsequently, above the certain dimension beam gives out under its own weight.

A similar reasoning leads us to understand why the animals can't be too big. The bones of legs of the hypothetical horse 10 times bigger as compared to the real ones would crack under their own heavyweights. It is now known that the main reason for that is that the substances are built of atoms and molecules, which have a definite size. Certainly, we can't build one of above-mentioned apparatuses so small in order to be made of the few molecules (Harvey, 1990; Costello et al., 2009).

As another example, assume the heavenly bodies. The stars and largest planet release light, the smaller planet, such as earth, do not. Only if the size, henceforth the mass of a body, is large enough, the temperature and pressure in its core, which is because of the action of gravitational forces amongst its parts, are large to trigger the thermonuclear fusion reactions that are able to produce light (Debunne et al., 2001; Swanton, 2012).

3.4.1. Mathematical formalism of Physical Laws

The laws of physics are not invariant under the changes of scale.

Now come to the mathematics of the reference frames. In following another type of, comparable, co-ordinates, *spherical polar coordinates need to be used*. Figure 3.2a displays such co-ordinates on plane, Figure. 3.2b in the space, together with orthogonal coordinates in both of the cases. On the plane, two polar co-ordinates of generic point P are the distance from origin ρ, which is the non-negative number, known as *radius*, and angle ϕ, amongst the x-axis and segment OP, known as *azimuth*. It is measured in the anticlockwise direction and changes between 0 and 2π.

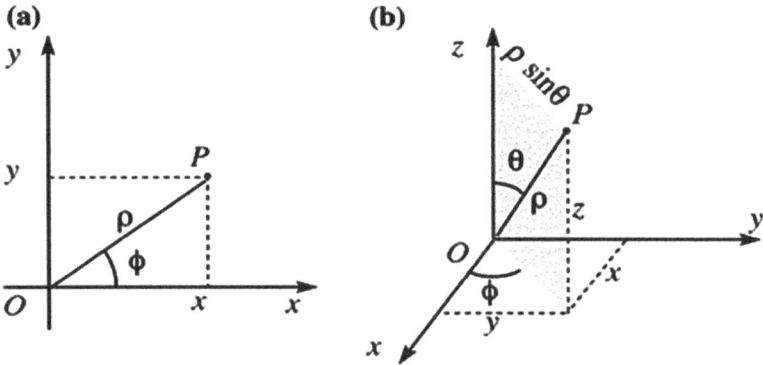

Figure 3.2. Polar coordinates: (a) 2 dimensions, (b) 3 dimensions.

Source: http://www.softouch.on.ca/kb/data/Course%20in%20Classical%20
Physics%201--Mechanics%20(A).pdf.

$$\rho \geq 0, \quad 0 \leq \phi < 2\pi \tag{3}$$

It can be easily seen from Figure 3.2 that the relations amongst polar and the rectangular co-ordinates are:

$$x = \rho \sin \theta \cos \phi, \quad y = \rho \sin \theta \sin \phi, \quad y = \rho \cos \theta \tag{4}$$

and inverse relations:

$$\rho = \sqrt{x^2 + y^2}, \quad \phi = \arctan \frac{y}{x} \tag{5}$$

Figure 3.2b displays the polar co-ordinates in 3 dimensions. The first coordinate of generic point P is the distance r from the origin, the second co-ordinate is angle ϕ amongst the plane through z and P and plane xz, the third co-ordinate is angle θ amongst the segment OP and z-axis. Again, r is the non-negative number. The angle θ changes from 0 to π, in a way the semi-plane displayed in Figure 3.2.

The semi-plane rotates around the z when ϕ changes between 0 and 2π.

Hence, the relations with orthogonal co-ordinates areas:

$$\rho \geq 0, \quad 0 \leq \theta \leq \pi, \quad 0 \leq \phi < 2\pi. \tag{6}$$

$$x = \rho \sin \theta \cos \phi, \quad y = \rho \sin \theta \sin \phi, \quad y = \rho \cos \theta \tag{7}$$

and inverse ones:

$$\rho = \sqrt{x^2 + y^2 + z^2}, \ \phi = \arctan\frac{y}{x}, \theta = \arctan\frac{\sqrt{x^2 + y^2}}{z} \tag{8}$$

If point P is on xy plane, namely if the $\theta = 0$, Equation (8) becomes:
$x = \rho \cos\phi, \ y = \rho \sin\phi,$

which are similar to Equation (3.4).

To know the motion of the body one must know its position in the different instants. More accurately, intervals of time are measured, instead of absolute time. In practice a certain instant is chosen and is defined as *origin of the times*, for which the $t = 0$. Next, a time interval is chosen and defined as unit of time. In SI, it is second. In principle, one should also select one of the two orientations as positive, but the selection is obvious. It is imposed by Nature: positive direction of the time is from the past to the future. Accordingly, time of a happening is negative if it occurred before $t = 0$, which is positive if it occurred after that.

One must now ask: is the origin of the times arbitrary? As always, one must turn into an experiment. Let's go to one of the experimental apparatuses and let's repeat the experiment beginning from same initial state, for instance in morning, then an afternoon, and again in night, etc. For each trial origin of the time is taken as initial instant. It is observed that once all of the spurious elements are handled all the experiments progress in the same way. The origin of the times is random, time is homogeneous. The laws of physics are invariant under the translations in time. Additionally, similarly to space, no ultimate time interval exists.

It is said that the selection of positive direction of the time is imposed by Nature. Numerous books have outlined this issue, the "arrow of time." Here it is only stated that in the virtuously mechanical situation no arrow of the time exists. Suppose a billiard ball is hot and a movie is made of the motion of billiard ball hitting the other balls and the walls. If the movie is played backward, one can observe the perfectively legitimate evolution. One can't know if it is forwards or backward.

3.5. VECTORS

Numerous physical quantities, like atmospheric pressure and temperature, are represented by a single number. This isn't the case pf quantities like acceleration, velocity, or force. For instance, to determine the velocity of a car isn't enough to know how fast the car moves, is also necessary for a fully description to know in which particular direction is moving (Karasuyama and Melchers, 1988; Johansson, 2009).

Another example is the displacement in space. To know it one need must know how long the displacement is and in which particular direction it happens. These kinds of physical quantities are characterized by vectors (Truesdell, 1953; Croyle et al., 2001).

A vector is a mathematical entity. To define it, let's begin considering the line segments. A segment is known as *oriented* if one of the two senses is selected as positive. Two oriented segments are equipollent if they have some sense, the same length, and the same direction. A *vector is class of all of the oriented segments equipollent to the given one.* It is graphically characterized by an arrow. It is characterized by length, called the *magnitude, sense,* and *direction.* Differently from oriented segment, it isn't indicated by its position. The velocities of the two cars moving at around100 km/h heading towards West, one near to Paris, one near to London are the same (Philippsen et al., 1978; Reschel et al., 2002).

Once the reference frame is selected, one can represent the vector by an *ordered triple of the real numbers,* which are the *components* of vector in that reference. However, the well-ordered triple of the real numbers isn't necessarily the vector. To be so following vital property should be satisfied. Indeed, if one changes the reference frame, for instance rotating the axes, the components of vector change, but itself the vector does not (Cline, 1985; Coutinho et al., 2004). Vector is the definite object; the components of vector are the way to observe it in one or more frames. To fulfill these properties, vector components, the well-ordered triples in two frames, need to be connected by the well-defined relations, which will be now found (Benton et al., 2002; Bandaranayake et al., 2011).

Figure 3.3 exhibits the reference frame and point P of the co-ordinates x, y, z. Take into account the oriented segment from origin O to P and corresponding vector **r**. It is known as the *position vector,* and is, one can say, prototype of the vectors. Its components in a Cartesian reference frame which are considered are simply coordinates of the P, for example, the ordered triple (x, y, z). Let's now consider another reference with same origin and the axes rotated by an angle θ. The point P doesn't move and

r doesn't change. But its components x', y', z' are different. The general association among the two triplets is complex. For simplicity, two frames with same origin and the same z-axis are considered, as shown in Figure 3.4.

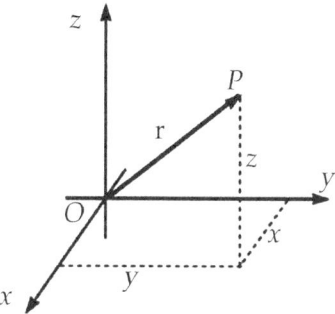

Figure 3.3. The orthogonal co-ordinates and position vector.

Source: http://www.softouch.on.ca/kb/data/Course%20in%20Classical%20 Physics%201--Mechanics%20(A).

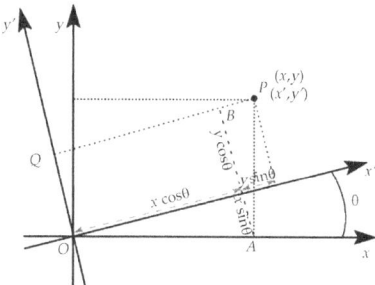

Figure 3.4. *Rotation of the Cartesian reference frame around common z-axis.*

Source: http://www.softouch.on.ca/kb/data/Course%20in%20Classical%20 Physics%201--Mechanics%20(A).pdf.

Let's take the point P in the figure of the co-ordinates x, y in one frame, x', y' in the other frame. One should express x' and y' as the functions of x, y, and the θ. One relation is clear, $z' = z$. In practice, one is reduced to just two dimensions (Salmon et al., 2001; Diebel, 2006).

Now perpendiculars are drawn from the P to all axes. The segment AB *is also drawn a* perpendicular to the PQ. The figure exhibits that x' is sum of the two lengths along x' and the y' *axis* difference of the two lengths along the *AB*. The following is obtained:

$$x' = x\cos\theta + y\sin\theta$$
$$y' = -x\sin\theta + y\cos\theta$$
$$z' = z, \qquad\qquad\qquad (9)$$

Where, in order to be complete, the third co-ordinate is also included. Notice that these associations are both relations amongst the components of position vector in two frames and the associations amongst the co-ordinates in two frames. As a matter of fact, they analytically describe the rotation of axes (Kelly et al., 2001; Palmer and Ng, 2004).

It can now be stated that the vector is a well-ordered triple of the real numbers that under the rotations of reference frame change in the same way as the triple signifying the position vector, namely the as co-ordinates.

Figure 3.5a signifies, in the plane for simplicity, the generic vector **A**, which can be thought as of drawn beginning from the origin, as all of the equipollent segments are same vector and its components in two frames (Clerk-Maxwell, 1869; Cheng and Zhang, 2009).

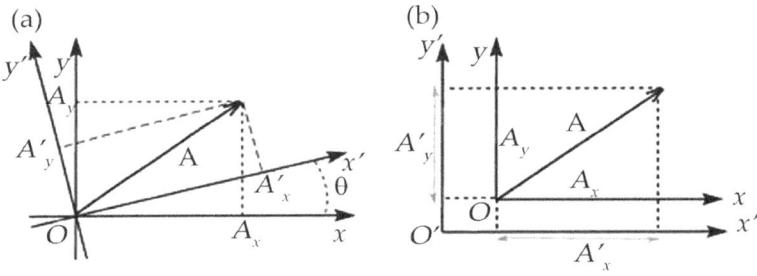

Figure 3.5. Components of vector A in the two frames distinct for (a) rotation (b) translation.

Source: http://www.softouch.on.ca/kb/data/Course%20in%20Classical%20 Physics%201--Mechanics%20(A).pdf.

By definition, the associations between its components are equivalent to Equation (9), namely:

$$A'_x + B'_x = (A_x + B_x)\cos\theta + (A_y + B_y)\sin\theta$$
$$A'_y + B'_y = -(A_x + B_x)\sin\theta + (A_y + B_y)\cos\theta$$
$$A'_z + B'_z = A_z + B_z, \tag{10}$$

The inverse relations, the expressions of $A'x$, $A'y$, $A'z$ as the functions of $A x$, $A y$, $A z$, and θ can be attained in two methods: inverting the system or, which is quite simpler, assuming that first reference is acquired from the second by the rotation of an angle $-\theta$. Accordingly, we have:

$$A_x = A'_x \cos\theta + A'_y \sin\theta$$
$$A_y = -A'_x \sin\theta + A'_y \cos\theta$$
$$A_z = A'_z, \tag{11}$$

Two frames are considered differing by the rotation of axes, with the common origin. Consider two frames differing for the translation, with parallel axes and the different origins, as displayed in Figure 3.5b, again for the ease in the plane. It can be seen that the components of vector **A** in two frames are equal (Winans, 1977; Pavlačka and Talašová, 2010).

3.6. OPERATIONS WITH VECTORS

The quantity characterized by the number, like pressure or temperature, is known as *scalar*. Scalars are invariant under the rotations of axes. In the two reference frames, rotated one to other the scalar has same value. Observe that every quantity is not scalar. For instance, x component of the vector isn't, as it changes under the rotations (Akutsu et al., 1992; Weiskopf et al., 2001).

A vector is represented with its components in the given frame with **A** = A x, A y, A z.

Provided the vector **A** and scalar k their product is k **A** = kA x, kA y, kA z. The components of the vector k **A** are simply k times those of the vector **A**. To be definite, it must be verified that the given definition agrees with definition of the vector. Indeed, it is instantaneous to validate that the oriented triple kA x, kA y, kA z transforms like a vector (Bauchau and Choi, 2003; Ge et al., 2004).

Geometrically, $k\mathbf{A}$ is a vector having same direction as the vector $\mathbf{A}$, the magnitude k times than that vector of $\mathbf{A}$ and sense of the $\mathbf{A}$ or opposite which is dependent on k being negative or positive respectively.

The product of vector $\mathbf{A}$ times the reciprocal of the magnitude is the vector with the direction of the $\mathbf{A}$ and unitary magnitude. A vector having unitary magnitude is known as *unit vector* or the *versor*. The symbol $\mathbf{u}$ A will be used for unit vector.

The product of vector $\mathbf{A}$ and -1 is known as *opposite* of the $\mathbf{A}$. It has same direction and magnitude of $\mathbf{A}$ and the opposite sense.

Consider two vectors A and B, which in the given reference frame have components $A\,x, A\,y, A\,z$ and $B\,x, B\,y, B\,z$ respectively. Consider triple of the numbers that are sums of homologous components of the A and B, $A\,x + B\,x, A\,y + B\,y, A\,z + B\,z$. Is it the vector? Let's check. Knowing that A and B vectors it is known that:

$$A_x' = A_x \cos\theta + A_y \sin\theta \qquad B_x' = B_x \cos\theta + B_y \sin\theta$$
$$A_y' = -A_x \sin\theta + A_y \cos\theta \qquad B_y' = -B_x \sin\theta + B_y \cos\theta$$
$$A_z' = A_z, \qquad\qquad\qquad B_z' = B_z,$$

By summing component to component we have:

$$A_x'B_x' + A_y' + B_y' + A_z' + B_z' = (A_x + B_x)\cos\theta + (A_y + B_y)\sin\theta$$
$$= -(A_x + B_x)\sin\theta + (A_y + B_y)\cos\theta$$
$$= A_z + B_z,$$
$$A_x'B_x' + A_y' + B_y' + A_z' + B_z' = (A_x \cos\theta + A_y \sin\theta) + (B_x \cos\theta + B_y \sin\theta) + (-A_x \sin\theta + B_x)\sin\theta)$$

It is seen that the answer is positive. It can then be defined as vector sum of the two vectors the vector having components equal to sums of their homologous components. Observe that the just discovered properties are instantaneous consequences of component transformations being the linear operations (Li and Hodgson, 2004; Deaton et al., 2005).

It is important to confirm that the sum of the vectors has common properties of the sum, commutative;

$$A + B = B + A \qquad\qquad\qquad (12)$$

and associative;

$$(A + B) + C = A + (B + C) \qquad\qquad\qquad (13)$$

Figure 3.6 exhibits the geometric meaning of the vector sum. In Figure 3.6a sum is built putting the tail of vector **B** on head of the vector **A**; the sum is vector from tail of the **A** to head of the **B**, as one instantaneously understands thinking to components. For commutative property, it may have done vice versa, starting from **B** and then putting the tail of vector **A** on head of the **B**.

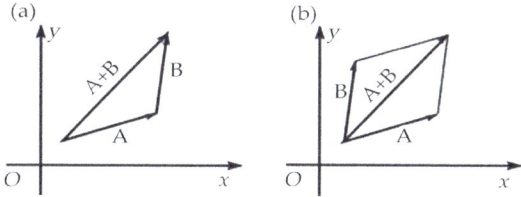

Figure 3.6. The sum of the two vectors.

Source: http://www.softouch.on.ca/kb/data/Course%20in%20Classical%20 Physics%201--Mechanics%20(A).pdf.

Figure 3.6b exhibits an equivalent way to the sum, the parallelogram rule. Both vectors are put with the tails in same point and a parallelogram is drawn having them as the sides.

The vector difference between the two vectors **a** and **b** is a vector of the components equal to differences amongst the homologous components or, equally, sum of *a* and *b* (i.e., x). The geometrical meaning is displayed in Figure 3.7.

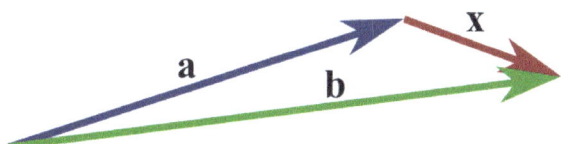

Figure 3.7. The difference between the two vectors.

Source: https://mathinsight.org/vector_introduction.

The properties of the vector sums, or configuration, which have been just discussed seems to be clear, but they aren't. Indeed, they are used when space is flat, not if space has any curvature. To make things easier, consider two dimensions. The plane surface is generally flat, but not the spherical one or the saddle-shaped one. The surface of earth on which we reside is flat only if the distances are considered substantially smaller as compared to the radius of earth, which has the mean value of $R = 6\,371$ km.

Let's consider the example of the vector addition. Consider the vector with tail in A at 45° in latitude and 0° in longitude and head B on same meridian at the 46° latitude. The length of 1° along the meridian is everywhere 10 thousand km/90 = 111 km. The second vector has tail in the B and the head on same parallel hundred km towards the West, say in C. Now the operations are commuted. We start with the vector hundred km long from A to D on its parallel at a hundred km to the West. Then a 111 km long vector is added to North with the tail in D and the head in C'. Is C' being equal to the C? The answer is NO. This is as the distance amongst two meridians along the parallel is different at the different latitudes.

The question of whether or not space has curvature must be answered experimentally. All measurements are well-matched with the zero mean curvature, inside their uncertainties meaning that a least from a local point of view the space is flat. However, it should be mentioned that space curvature prevails in the context of General relativity, where it defines the local effects of gravity in terms of modification of the geometry in the space surrounding the massive object.

3.7. SCALAR PRODUCT OF THE TWO VECTORS

There are two methods to obtain a product of the two vectors, called *cross product* and *dot product* respectively (Bracken et al., 1991; Goethals et al., 2004). Consider two vectors **A** and **B**. The dot product of these vectors is indicated with the dot between them, **A.B**. In the given reference frame, the dot product is the sum of products of the homologous components.

$$A.B = A_x B_x + A_y B_y + A_z B_z, \qquad (14)$$

The dot product possesses a vital property to be scalar, invariant under the rotations of axes. It is subsequently also known as the *scalar product*. Let's exhibit the property, that;

$$A'_x B'_x + A'_y B'_y + A'_z B_z = A_x B_x + A_y B_y + A_z B_z,$$

For simplicity, let's consider only the rotation around the *z-axis*. The components of the vector **A** in rotated frame as the functions of its components in starting one are usually given by Equation (3.10) and likewise for **B**. It can be written as:

$$A_x'B_x' + A_y' + B_y' + A_z' + B_z' = (A_x \cos\theta + A_y \sin\theta) + (B_x \cos\theta + B_y \sin\theta)$$
$$+ (-A_x \sin\theta + A_y \cos\theta)(-B_x \sin\theta + B_y \cos\theta) + A_z B_z$$
$$= A_x B_x \cos^2\theta + A_x B_y \cos\theta \sin\theta + A_y B_y \cos\theta \sin\theta$$
$$+ A_y B_y \sin^2\theta + A_x B_x \sin^2\theta - A_y B_x \sin\theta \cos\theta$$
$$- A_y B_x \cos\theta \sin\theta + A_y B_y \cos^2\theta + A_z B_z$$
$$= A_x B_x + A_y B_y + A_z B_z$$

It is observed that the product is invariant under such a transformation. It is simple to exhibit that both the distributive and commutative properties are true for the dot product.

$$A . (B + C) = A . B + A . C \tag{15}$$

Lets now see the geometric meaning of the scalar product. We can benefit from dot product being invariant to select convenient axes. X is taken in direction of the vector **A** and y in-plane defined by the **A** and **B**. If θ is angle amongst the vectors, then components are $A = A, 0, 0$ and $B = B\cos\theta$, $B\sin\theta$, 0. The dot product is given as:

$$\mathbf{A}.\,\mathbf{B} = AB\cos\theta \tag{16}$$

In words, the scalar product of the two vectors is the product of the magnitude times cosine of the angle between them. There are two other understandings that might be useful. The scalar product is product of magnitude of first vector times the projection of second vector on first one (Figure 3.8b), or, same with the inverted roles (Figure 3.8c).

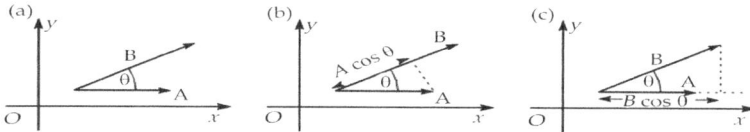

Figure 3.8. Geometric meanings of the scalar product.

Source: http://www.softouch.on.ca/kb/data/Course%20in%20Classical%20 Physics%201--Mechanics%20(A).

The dot product is 0 if vectors are perpendicular, negative if the angle is obtuse, and positive if acute.

An interesting and particular case is a product of the vector by itself;

$$A.A = A_x^2 + A_y^2 + A_z^2 = A^2.$$
(17)

By definition square of the vector is dot product of vector times itself and is equivalent to the square of the magnitude and also to sum of squares of the components. The last property is an instantaneous result of the Pythagorean theorem. It is also known as the *norm* of vector. The norm is clearly the same in all references (Sazdjian, 1988; Amirbekyan and Estivill-Castro, 2007).

Figure 3.9 exhibits the Cartesian reference frame where three significant vectors are drawn, unit vectors of coordinate axes, **i, j,** and **k**. They possess unit magnitude and are thus mutually normal. Consequently,

$i . i = 1, j . j = 1, k . k = 1$

$i . j = 0, j . k = 0, k . i = 0.$ (18)

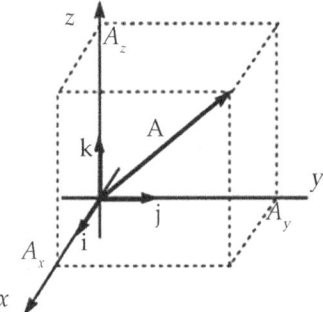

Figure 3.9. The unit vectors of Cartesian axes.

Source: http://www.softouch.on.ca/kb/data/Course%20in%20Classical%20 Physics%201--Mechanics%20(A).

The components of a vector can be outlined in terms of three unit vectors. Certainly, the x component of vector **A** is the dot product with **i**, since the magnitude of latter is 1, and likewise for other components (Mackey et al., 2004; Noui and Perez, 2005; Yang et al., 2006). The vector can then be written as:

$A = A_x i + A_y j + A_z k,$ (19)

Namely, as a sum of the three vectors having directions of the axes. These are known as vector components (Bechtel et al., 1971; Knyazev and Argentati, 2002). Particularly position vector can be described as:

$r = xi + yj + zk.$ (20)

$$C = C_x i + C_y j + C_z k = A \times B$$
$$= (A_y B_z - A_z B_y)i = (A_z B_x - A_x B_z)j = (A_x B_y - A_y B_x)k. \qquad (21)$$

3.8. VECTOR PRODUCT OF THE TWO VECTORS

Given two vectors $\mathbf{A} = A\,x,\ A\,y,\ A\,z$ and $\mathbf{B} = B\,x,\ B\,y,\ B\,z$, then the cross product is described as ordered triple of the real numbers (Williams and Stein, 1964; Silva and de Andrade Martins, 2002). The cross product changes as the vector under rotations of axes and is also known as the *vector product*. For the x' component, the presentation for other two is precisely the same.

$$C'_x = (A \times B)_x = A'_y B'_z - A'_z B'_y = (-A_x \sin\theta + A_y \cos\theta)\,B_z - A_z\,(-B_x \sin\theta + B_y \cos\theta)$$

$$= (A_z B_x - A_x B_z)\sin\theta + (A_y B_z - A_z B_y)\cos\theta + C_y \sin\theta + C_x \cos\theta$$

The vector product isn't commutative, and the order of factors matters.

$$B \times A = -A \times B. \qquad (22)$$

It can be understood from definition that:

Inverting the order of factors the sign of product changes. The property is termed as anticommutative.

It is simple to observe that the vector product is usually distributive to sum.

$$A \times (B + C) = A \times B + A \times C. \qquad (23)$$

Now the geometric meaning of cross product is observed using the same frame as in the previous section. Two vectors are drawn as starting from same point and then take the x-axis in direction and the sense of $\mathbf{A}$, the y-axis in the plane of two vectors and the z-axis in order to comprehend the right-handed reference (Figure 3.10). The components are $\mathbf{A} = A, 0, 0$ and $\mathbf{B} = (B\cos\theta, B\sin\theta, 0)$. The cross product possesses only the z-component different from 0.

$$A \times B = kAB\sin\theta \qquad (24)$$

VECTOR PRODUCT ("CROSS" PRODUCT)

The vector product of $\vec{A}$ and $\vec{B}$, written as $\vec{A} \times \vec{B}$, produces a third vector $\vec{C}$ whose magnitude is

$$\vec{C} = \vec{A} \times \vec{B} = |\vec{A}||\vec{B}|\sin\theta\,\hat{n} \qquad\qquad -\vec{C} = \vec{B} \times \vec{A}$$

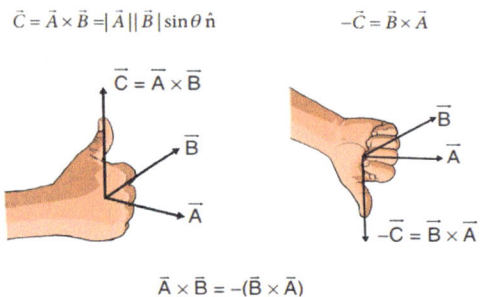

$$\bar{A} \times \bar{B} = -(\bar{B} \times \bar{A})$$

Figure 3.10. The vector product of the two vectors.

Source: https://www.brainkart.com/article/The-Vector-Product-of-Two-Vectors_34464/.

Therefore, the cross product is generally in a positive direction of the z-axis if the angle θ is in one of first two quadrants (Figure 3.10a), in negative direction if in the fourth and third ones (Figure 3.10b).

The conclusion, the geometric meaning of vector product, independently of the reference frame, is as follows. The magnitude of vector product is equal to area of the parallelogram possessing two vectors as the sides (Stetson, 1975; Gray and Rolland, 2015). Alternatively, it can also be said that the magnitude of vector product is the magnitude of first (A) times projection of second on the normal to first ($B \sin \theta$) or vice versa. The direction of vector product is perpendicular to plane of the two vectors.

Notice the same convention is followed here as used to describe the positive direction of the z-axis. In the left-handed frame, the sense of vector product might have changed too (Mallik, 2008; Widnall, 2009).

$$i \times j = k, j \times k = i, k \times i = j. \tag{25}$$

The cross product is 0 if any of the two vectors is 0 or if both are parallel. Particularly the product of vector times itself is 0. Every unit vector of the axes is a cross product of other two.

Let's describe *scalar triple product* of the three vectors, in order **A**, **B**, and **C**. It is dot product of first vectors times cross product of second vector times the third:

$$A.(B \times C) = A_x(B_yC_z - B_zC_y) + A_y(B_zC_x - B_xC_z) + A_z(B_xC_y - B_yC_x) \quad (26)$$

To understand the geometrical meaning, three are taken vectors beginning from the same point as in Figure 3.11.

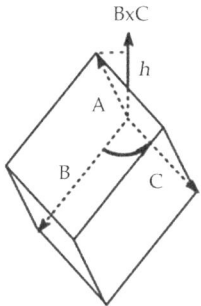

Figure 3.11. Scalar triple product.

Source: http://www.softouch.on.ca/kb/data/Course%20in%20Classical%20 Physics%201--Mechanics%20(A).

Let's consider them as sides of the parallelepiped. As the magnitude of the **B** × **C** is equal to area of the parallelogram possessing two vectors as the sides, which is the face of parallelepiped. Its direction is normal to that specific plane and positive sense is one that observes **B** going to the **C**, rotating through smaller angle, in the anticlockwise direction. Let suppose that the **A** lies on same side of plane made by the **B** and C as **B** × **C**. The dot product of vector **A** times **B** × **C** is product of projection of the **A** on direction of the **B** × **C** therefore on the direction perpendicular to plane of the **B** and C times magnitude of the **B** × **C**. This projection is only the height of the parallelepiped. The triple product is thus equal to volume of the parallelepiped possessing three vectors as the sides. In this situation, it is considered that this is valid in absolute sign and value (Gessler, 1973).

The subsequent properties are immediately demonstrated: triple scalar product is 0 if all 3 vectors are coplanar, therefore, particularly, if 2 or 3 are parallel. The triple product doesn't change if factors are circularly permuted. Obviously also;

$$A.(A \times C) = 0 \quad\quad\quad\quad (28)$$

The second triple product is a *triple vector product*, which normally is the cross product of first vector time's cross product of second and the third ones. By the direct verification one exhibits that;

A. (B × C) = B (A . C) – C(A . B)   (29)

3.9. BOUND VECTORS, COUPLE, MOMENT

The forces are the vectors. However, to fully characterize the force one also need to find out its *application point*. If an object is pushed with the finger, on it not only an action of certain intensity and in a certain direction is exerted, but also exerted in a certain point (Scott, 1971; Leurer, 1994). If that point is changed, the effect of force would also change. A vector with linked application point is known as the *bound vector* (Vasco, 1990; Marghitu, 2005).

The line with the orientation of the force through the application point is known as the line of action.

Figure 3.12 exhibits vector **A** and the point of application *P*. Arbitrarily a point Ω is chosen which is called the *pole*. The *moment* of vector **A** about the Ω is described as the vector product of vector leading from the pole to application point of the **A**, namely;

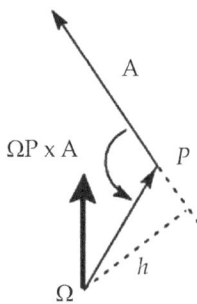

Figure 3.12. Moment of the vector A about pole Ω.

Source: http://www.softouch.on.ca/kb/data/Course%20in%20Classical%20 Physics%201--Mechanics%20(A).

τΩ = ΩP × A.   (30)

Let's examine its geometrical meaning. The direction of moment of the **A** is perpendicular to plane described by segment ΩP and the **A**. To observe its positive direction imagine vector **A** to be the force and ΩP the rigid bar. If it is observed that the force is turning the bar in the anticlockwise direction, then it's the positive side of moment (Inoue et al., 1986; Couture and König,

1996; Close and Törnqvist, 2002). The magnitude of moment is defined by-product of the magnitude of distance of the pole Ω from the action line of the **A**. In particular, if the pole Ω lies on action line then the moment is zero. Now consider an easier and particularly vital situation, the *couple* vectors. A couple is the pair of the bound vectors which are equal in magnitude and are in opposite direction. The distance amongst the two action lines is known as the *arm* of couple (Casalbuoni et al., 1989; Tagliani, 2003).

A very significant property of the couple is that the moment is independent of pole. This may be known as the *moment of couple* or the *couple torque* (Balazs et al., 1999; Hong and Kim, 2009).

The 2 terms are synonymous. Consider for easiness the pole lying in the plane of the couple, as shown in Figure 3.13. The two vectors are -**A** and **A**. *P* 1 and *P* 2 are the application points. The total moment i.e., the sum of two moments about the pole Ω is:

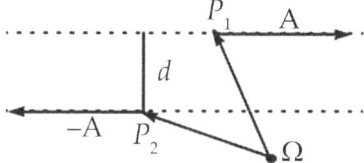

Figure 3.13. A couple of the bound vectors.

$$\tau\Omega = \Omega P_1 \times A - \Omega P_2 \times A = (\Omega P_1 - \Omega P_2) \times A = P_2 P_1 \times A, \qquad (31)$$

which is not dependent on the pole. It can also be seen that the magnitude of couple moment is the product of magnitude A of vectors times arm d of the couple;

$$\tau = d \cdot A. \qquad (32)$$

The direction of the couple moment is perpendicular to plane of the couple; positive on side viewing the couple rotates in the anticlockwise direction (Lipkin and Duffy, 1982; Tsiatas, 2009).

3.10. VELOCITY

Let's understand the motion of simplest body, a *particle,* or *material.* Let us consider the circumstance when its dimensions are quite small compared to the distances from the other objects. This is obviously an idealization but still it works frequently in practice (Alder and Wainwright, 1970; Bussi et al., 2007). For instance, the planets are indeed not point-like, but in a mathematical description the motions of planets around the sun can be well-thought-out as such in good estimate, as long as one doesn't take into account rotations about their axes, or changes of the directions of axes, or tides on the surfaces (Da Rold and Pomarol, 2006; Spurr, 2008). The ship can be thought as a point when it is very far from the shore, but when ship enters the harbor its dimension should be precisely known (Tanner and Whitehouse, 1976; Loarie et al., 2009).

As already defined, the motion is to be examined in the given reference frame. The particle defines in its motion the curve, which is known as the *trajectory,* as displayed in Figure 3.14a. The position vector is the function of time $\mathbf{r}(t)$, in other words, co-ordinates are the 3 functions of time $x(t)$, $y(t)$, $z(t)$. If these functions are known, the motion of particle is completely known. It can be said that the system possesses $3°$ of freedom.

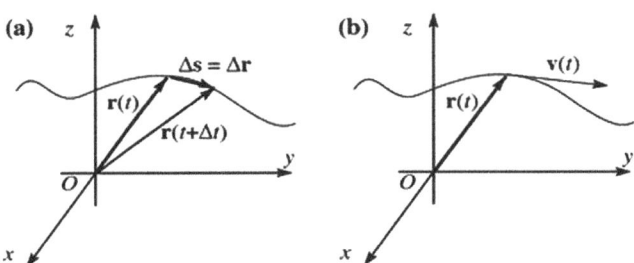

Figure 3.14. (a) The trajectory of the particle, (b) velocity.

Source: http://www.softouch.on.ca/kb/data/Course%20in%20Classical%20 Physics%201--Mechanics%20(A).

Let's consider position vector at instant of the time t, $\mathbf{r}(t)$ as signified in Figure 3.14a and an instantaneously following instant $t + \Delta t$, $\mathbf{r}(t + \Delta t)$, Δt is the short time interval. In this short time interval the particle has progressed by $\Delta\,\mathbf{s}$, which is the step in space having the magnitude and direction, it is the vector (Alder and Wainwright, 1967; Perrine and Edgerton, 1978).

Looking at figure one instantaneously observes that the Δ **s** is equal to difference amongst the 2 vectors **r**(t + Δt) and **r**(t). This is the change of vector **r** in time interval Δt. Hence,

$$\Delta s = \Delta r = r(t + \Delta t) - r(t). \tag{33}$$

Average velocity in time interval Δt is vector attained by dividing the displacement with the time interval in which the displacement happens:

$$\langle v \rangle = \frac{\Delta s}{\Delta t} = \frac{\Delta r}{\Delta t} = \frac{r(t + \Delta t) - r(t)}{\Delta t} \tag{34}$$

or, in components;

$$\langle v_x \rangle = \frac{x(t + \Delta t) - x(t)}{\Delta t}, \langle v_y \rangle = \frac{y(t + \Delta t) - y(t)}{\Delta t}, \langle v_z \rangle = \frac{z(t + \Delta t) - z(t)}{\Delta t} \tag{35}$$

Velocity is the boundary or restriction for average velocity, namely:

$$v = \frac{dr}{dt}. \tag{36}$$

In words, the velocity is a time derivative of position vector. Its components are derivatives of the coordinates:

$$v_x = \frac{dx}{dt}, v_y = \frac{dy}{dt}, v_z = \frac{dz}{dt}. \tag{37}$$

In the limit, the orientation of Δ **s** gets tangent to trajectory, in all points of trajectory the orientation of the velocity is that of tangent in that specific point. The physical dimensions of the velocity are those of length divided by the time; the unit is subsequently meter per second (m/s) (Salomone et al., 1981; Buckley et al., 2015; Ghorbani, 2015).

The motion is *uniform* if the magnitude of the velocity doesn't vary in time. In the uniform motion though, the velocity isn't necessarily constant, as its direction might vary. The direction of the velocity doesn't change if motion is rectilinear. Hence, motion with the constant velocity is the *rectilinear uniform* (Taner and Koehler, 1969; McDicken et al., 1992).

A vital motion is circular one, in this motion, the trajectory is a circle. Let R be the radius. It is always suitable to select the reference frame taking benefit from symmetry of the issue if any is existent. Here the origin is taken in the center of circle and the z-axis perpendicular to the plane. The motion is in the xy plane as displayed in Figure 3.15a (Harlander and Kilgore, 2002; Lahanas and Spanos, 2002).

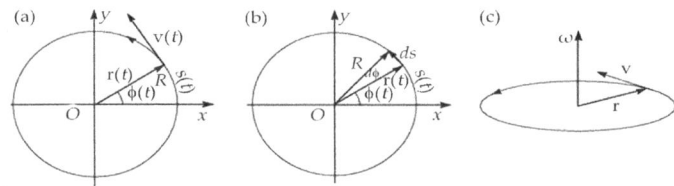

Figure 3.15. *(a) Circular motion, (b) infinitesimal movement, (c) ω (angular velocity).*

Source: http://www.softouch.on.ca/kb/data/Course%20in%20Classical%20 Physics%201--Mechanics%20(A).

Further, the origin of the time is chosen in moment in which point crosses the positive x-axis. Let $\phi(t)$ be an angle amongst the x-axis and position vector at the time t, considered as positive in an anticlockwise direction and assume s(t) be the length of arc subtended by the $\phi(t)$, taken with same sign as the ϕ, $s(t) = R\,\phi(t)$. Let ds be infinitesimal movement in the dt (Figure 3.15b). The infinitesimal variations of ϕ and s are associated by relation $ds = R\,d\phi$, where ds is magnitude of the ds if motion is anticlockwise (Figure 3.15), and is opposite if the motion is clockwise (Hursh, 1939; Kraichnan, 1970; Van Beveren et al., 1986; Durrer et al., 1998). The angular velocity measures rate of change of angle. Consider the time derivative:

$$\omega_z = \frac{d\phi}{dt}. \tag{38}$$

This quantity has a sign and magnitude, depending on the sense of the rotation. In fact, it's the z-component of angular velocity, which is the vector. Its magnitude is the absolute value of the Equation (46), its direction is usually perpendicular to plane of the motion, considered positive on side viewing the motion is anticlockwise (Miller and Phillips, 1969; Gnecco et al., 2000). This is the z-axis in Figure 3.15c.

The physical dimensions of angular velocity are inverse of the time; its unit is rad/s (radians per second).

In the circular motion, magnitudes of the velocity $\upsilon = |ds|/dt$ and magnitude of angular velocity ω are related by:

$$v = \omega R. \tag{39}$$

The relation amongst corresponding vectors, as instantaneously seen from Figure 3.15c is:

$v = \omega \times r.$ (40)

Let's consider the situation in which magnitude υ of velocity is constant.

The motion is uniform and circular, the arcs and conforming angles are proportional to times taken in order to travel them, $\phi(t) = s(t) / R = \pm vt / R$

Hence, equations of motion in the polar coordinate:

$$r(t) = R, \theta(t) = \frac{\pi}{2}, \phi(t) = \pm vt / R = \omega_z t$$
(41)

The equations of the motion in the Cartesian coordinates are:

$$x(t) = R \cos \omega_z t, y(t) = R \sin \omega_z t, z(t) = 0.$$ (42)

As an exercise, one can validate that the trajectory is the circle. Taking the squares of members and then summing the following equation of the circumference is obtained (Lighthill, 1970; Sexton et al., 1995; He and Yuan, 1999).

$$x^2(t) + y^2(t) + z^2(t) = R^2.$$ (43)

Notice that two Cartesian coordinates x and y aren't independent but if one is known the other can also be known. The particle is assumed to move onto the prefixed trajectory (Foote and Du Toit, 1969; Wakker and van Woerden, 1997). The system has on degree of freedom. This is obvious in the polar co-ordinates, Equation (40).

Cartesian components of the velocity are expressed as:

$$v_x(t) = \frac{dx}{dt} = -\omega_z R \sin \omega_z t = v_y(t) = \frac{dy}{dt} = \omega_z R \cos \omega_z t, v_z(t) = \frac{dz}{dt} = 0$$
(44)

The components of the velocity vector vary in time: when particle travels on the circle, the direction of particle continuously changes even if the magnitude of particle is constant. Indeed, magnitude is:

$$v = \sqrt{v_x^2 + v_y^2 + v_z^2} = \omega R \sqrt{\cos^2 \omega_z t + \sin^2 \omega_z t} = \omega R$$ (45)

which is constant.

As an additional exercise, let's validate that velocity is continuously tangent to trajectory, i.e., perpendicular to position vector everywhere. To observe that their scalar product is taken and get:

$r(t).v(t)=x(t)v_x(t)+ u(t)v_y(t)=- \omega R^2 \cos\omega_z t \sin\omega_z t + \omega R^2 \sin\omega_z t \cos\omega_z t=0$

The following observation is made that will be helpful in the following. In the situation, it is noted that there are two vectors: the position and velocity vector. The x and y components of first vector which is position vector are proportional to sine and the cosine of angular co-ordinate respectively, whereas those of the second vector to opposite of its cosine and to its sine respectively (Hodgkin, 1954; Cremmer et al., 1978; Barnaby et al., 2011).

When this occurs, the vectors are perpendicular. Both the components of velocity and co-ordinates are proportional to circular functions $\sin\omega t$ or $\cos\omega t$, which are periodic. The motion is periodic if velocity and position have some values in instant t they again have the same values at instants $t + T$, $t + 2T$, etc. The time T is known as the *period* of motion. T is inversely proportional to angular velocity (Hecking et al., 1978; Hauser, 1986; Faraei and Jafari, 2017)

$$T = \frac{2\pi}{\omega}$$

(46)

3.11. ACCELERATION

The motion of the body in which velocity changes with time in direction or magnitude is called *acceleration*. If a change of the velocity in time interval Δt is given by $\Delta \mathbf{v}$, then average acceleration in that specific time interval is ratio (Blandford and Ostriker, 1978; Pulay, 1982; Jones and Ellison, 1991)

$$\langle a \rangle = \frac{\Delta v}{\Delta t}.$$

(47)

The instantaneous acceleration at the time t is the limit for $\Delta t \to 0$, the time derivative of velocity:

$$a = \frac{dv}{dt} \left[a_x = \frac{dv_x}{dt}, a_y = \frac{dv_y}{dt}, a_z = \frac{dv_z}{dt} \right]$$

(48)

In the specific case of rectilinear motion, when the orientation of velocity is constant, the direction of acceleration is also parallel and its sign and magnitude are:

$$a = \frac{dv}{dt}$$

(49)

Now consider the *uniform circular* motion where the velocity vector has constant magnitude and changes in direction with the constant angular velocity (Chenery, 1952; Kruger and Westermann, 2003). To find acceleration, take into account the auxiliary diagram of Figure 3.16a. The axes of the figure are x and y components of velocity vector. It is similar to the position vector in xy plane. The analogy is comprehensive as both vectors rotate with the constant angular velocity ω (Ashkin, 1970; Rideout and Breslow, 1980; Altshuler et al., 1995).

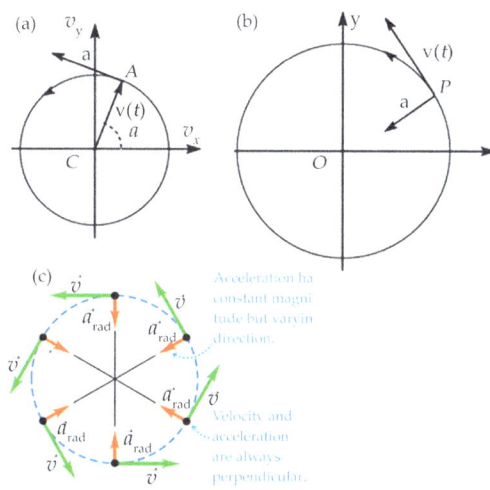

Figure 3.16. Uniform circular motion.

Source: https://physics.stackexchange.com/questions/38291/uniform-circular-motion.

Clearly, the velocity of point A is the acceleration of the particle P since the displacement of vector A in time interval dt is the d **v** and subsequently,

its velocity is $a = \dfrac{dv}{dt}$. The vector is tangent to circle and accordingly perpendicular to velocity (Figure 3.16a). More accurately, direction of the acceleration is attained from that of velocity by the rotation of 90° in the anticlockwise direction. Going back to demonstration of the motion in xy plane in Figure 3.16b, the acceleration, which must be at 90° from velocity anticlockwise, is directed towards the center. It is called *centripetal acceleration* (Luh et al., 1980; Bao and Intille, 2004).

The magnitude of the acceleration is immediately found. The angle between the vector **v** and abscissa axis in Figure 3.16a is denoted by α and *d*α its change in time *dt*. Considering that vector rotates with the constant angular velocity ω, $dα = ω$ dt. The relation for change in velocity is given

by $|dv| = vdα$ and an important relationship is obtained (Richtmyer, 1960; Bell, 1978; Weinberg et al., 2013).

$$a = \left|\frac{dv}{dt}\right| = v\frac{dα}{dt} = ωv = \frac{v^2}{R} = ω^2 R$$

(50)

3.12. MOTION ON THE PLANE

Consider the general movement in the plane. Unit vector tangent to trajectory in the generic point in orientation of the velocity in the point is indicated with **u** t. Generally, **u** t changes in time. Figure 3.17 exhibits the situation in two consecutive instants (Gutenberg and Richter, 1942; Polyak and Juditsky, 1992).

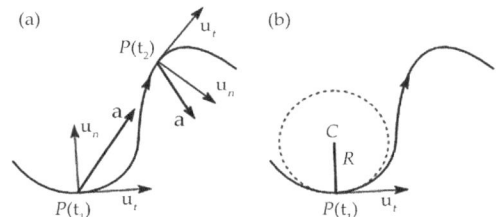

Figure 3.17. (a) The acceleration vector in the two different points of trajectory; (b) osculating circle.

Source: http://www.softouch.on.ca/kb/data/Course%20in%20Classical%20 Physics%201--Mechanics%20(A).

In every instant, for example. In every point of trajectory, generally the velocity is quite different. Unit vector normal to trajectory is indicated with **u** n. The positive direction is direction acquired by rotating unit vector by 90° in direction of the immediate rotation of velocity vector. This means that **u** n is directed towards curvature center (Fujiwara and Shimono, 1983; Gopalswamy et al., 2000). The latter might lie on right or the left of the trajectory dependent on the case. To attain the acceleration the derivative of velocity expressed as product of the magnitude times the unit vector is taken, **v** = υ **u** t.

$$\frac{dv}{dt} = \frac{dv}{dt}u_t + v\omega \cdot u_n = a_t u_t + a_n u_n$$

(51)

The acceleration possesses two components. One component is tangent to trajectory and is equal to the time derivative of magnitude of the velocity (Reames, 1999; Steffen et al., 2015). It is zero if the motion is uniform, positive if motion is accelerated, negative if motion is decelerated. The second component is normal to trajectory in any situation towards the "interior" of curve. It is zero when the direction of velocity doesn't change, even if immediately, as in flex points of trajectory (Sprangle et al., 1988; Pukhov et al., 1999).

The small curve segment can be approximated around the point P with an arc of the osculating circle and consider the point as moving on arc with the angular velocity $\omega = v/R$. In conclusion, two components of acceleration are as:

$$a_t = \frac{dv}{dt}, a_n = \frac{v^2}{R}.$$

(52)

3.13. VECTORS, PSEUDOVECTORS, SCALARS, AND PSEUDOSCALARS

Previously, the vector is defined as a well-ordered triple of the real numbers that under the rotations of reference frame changes in same way as the triplet characterizing the position vector (Kamiński et al., 1997; Maris and Roberts, 1998).

Previously, a scalar quantity is introduced, dot product of the two vectors. It is seen that scalar quantity is same in the two reference frames differing for the rotation of axes. Indeed, generally, a quantity, by definition, is scalar if the quantity is invariant under the change of reference frame (Hsiao et al., 2000; Chou, 2009).

Hence, both scalar and vector properties are articulated in terms of changes between the reference frames. Let's now contemplate the behaviors of both these quantities under inversion of the axes. It is known as a *parity* operation. It leads from the left-handed frame to right-handed one (McConnell, 1958; Bennhold and Wright, 1987).

Consider the variation properties of the physical quantities. The quantity can be scalar or can be pseudoscalar. Both are invariant under the rotations but the earlier is invariant under the parity operation, the latter varies sign, whereas keeping the absolute value (Deo and Bisoi, 1974; Chiu et al., 2006).

The dot product of the two vectors is scalar; the scalar triple product is pseudoscalar. This is immediately evident seeing that under the inversion of axes all the 3 vector factors transform sign. The quantity can be vector or pseudovector (Norton and Watson, 1958; Sugawara and Okubo, 1960). Both change in the same way under the rotations, but the components of earlier transform sign under the inversion of axes, whereas the components of latter don't change sign (Divotgey et al., 2013; Catà, 2014).

The cross product of the two vectors is pseudovector since both vector factors transform sign and their product doesn't. Position vector, acceleration, and velocity are vectors; moment and angular velocity of the vector are pseudovectors (Fushchich et al., 1989; Iwamoto, 1989).

These kinds of properties of physical quantities belong to the class generically known as *symmetry properties* (Glück, 1972; Noble, 1979; Li et al., 1998).

REFERENCES

1. Akutsu, A., Tonomura, Y., Hashimoto, H., & Ohba, Y., (1992). Video indexing using motion vectors. In: *Visual Communications and Image Processing'92* (Vol. 1818, pp. 1522–1530). International Society for Optics and Photonics.

2. Alder, B. J., & Wainwright, T. E., (1967). Velocity autocorrelations for hard spheres. *Physical Review Letters, 18*(23), 988.

3. Alder, B. J., & Wainwright, T. E., (1970). Decay of the velocity autocorrelation function. *Physical Review A, 1*(1), 18.

4. Alexander, S., (1920). *Space, Time, and Deity: the Gifford Lectures at Glasgow, 1916–1918* (Vol. 2, pp. 1–10). Macmillan.

5. Altshuler, L. L., Post, R. M., Leverich, G. S., Mikalauskas, K., Rosoff, A., & Ackerman, L., (1995). Antidepressant-induced mania and cycle acceleration: A controversy revisited. *The American Journal of Psychiatry, 152*(8), 1130–1138.

6. Amirbekyan, A., & Estivill-Castro, V., (2007). A new efficient privacy-preserving scalar product protocol. In: *Proceedings of the Sixth Australasian Conference on Data Mining and Analytics* (Vol. 70, pp. 209–214). Australian Computer Society, Inc.

7. Ashkin, A., (1970). Acceleration and trapping of particles by radiation pressure. *Physical Review Letters, 24*(4), 156.

8. Balazs, C., He, H. J., & Yuan, C. P., (1999). QCD corrections to scalar production via heavy quark fusion at hadron colliders. *Physical Review D, 60*(11), 114001.

9. Bandaranayake, A. D., Correnti, C., Ryu, B. Y., Brault, M., Strong, R. K., & Rawlings, D. J., (2011). Daedalus: A robust, turnkey platform for rapid production of decigram quantities of active recombinant proteins in human cell lines using novel lentiviral vectors. *Nucleic Acids Research, 39*(21), e143–e143.

10. Banuelos, R., & Méndez-Hernández, P. J., (2003). Space-time Brownian motion and the Beurling-Ahlfors transform. *Indiana University Mathematics Journal, 1*(1), 981–990.

11. Bao, L., & Intille, S. S., (2004). Activity recognition from user-annotated acceleration data. In: *International Conference on Pervasive Computing* (Vol. 1, pp. 1–17). Springer, Berlin, Heidelberg.

12. Barnaby, N., Namba, R., & Peloso, M., (2011). Phenomenology of a pseudo-scalar inflation: Naturally large nongaussianity. *Journal of*

Cosmology and Astroparticle Physics, 2011(04), 009.

13. Bauchau, O. A., & Choi, J. Y., (2003). The vector parameterization of motion. *Nonlinear Dynamics, 33*(2), 165–188.

14. Bechtel, G. G., Tucker, L. R., & Chang, W. C., (1971). A scalar product model for the multidimensional scaling of choice. *Psychometrika, 36*(4), 369–388.

15. Bell, A. R., (1978). The acceleration of cosmic rays in shock fronts–I. *Monthly Notices of the Royal Astronomical Society, 182*(2), 147–156.

16. Beltrametti, E. G., Cassinelli, G., & Lahti, P. J., (1990). Unitary measurements of discrete quantities in quantum mechanics. *Journal of Mathematical Physics, 31*(1), 91–98.

17. Bennhold, C., & Wright, L. E., (1987). Pseudovector versus pseudoscalar theory in kaon photoproduction from nucleons and nuclei. *Physical Review C, 36*(1), 438.

18. Benton, T., Chen, T., McEntee, M., Fox, B., King, D., Crombie, R., & Bebbington, C., (2002). The use of UCOE vectors in combination with a preadapted serum free, suspension cell line allows for rapid production of large quantities of protein. *Cytotechnology, 38*(1–3), 43–46.

19. Blandford, R. D., & Ostriker, J. P., (1978). Particle acceleration by astrophysical shocks. *The Astrophysical Journal, 221*, L29–L32.

20. Bracken, A. J., McAnally, D. S., Zhang, R. B., & Gould, M. D., (1991). A q-analog of Bargmann space and its scalar product. *Journal of Physics A: Mathematical and General, 24*(7), 1379.

21. Buckley, M. R., Feld, D., & Goncalves, D., (2015). Scalar simplified models for dark matter. *Physical Review D, 91*(1), 015017.

22. Busch, P., (1991). Informationally complete sets of physical quantities. *International Journal of Theoretical Physics, 30*(9), 1217–1227.

23. Bussi, G., Donadio, D., & Parrinello, M., (2007). Canonical sampling through velocity rescaling. *The Journal of Chemical Physics, 126*(1), 014101.

24. Casalbuoni, R., De Curtis, S., Dominici, D., Feruglio, F., & Gatto, R., (1989). Vector and axial-vector bound states from a strongly interacting electroweak sector. *International Journal of Modern Physics A, 4*(05), 1065–1110.

25. Cassinelli, G., & Lahti, P. J., (1989). The measurement statistics interpretation of quantum mechanics: Possible values and possible

measurement results of physical quantities. *Foundations of Physics*, *19*(7), 873–890.

26. Catà, O., (2014). Lurking pseudovectors below the TeV scale. *The European Physical Journal C*, *74*(8), 2991.

27. Chenery, H. B., (1952). Overcapacity and the acceleration principle. *Econometrica: Journal of the Econometric Society*, 1(1), 1–28.

28. Cheng, Z., & Zhang, X., (2009). Estimating differential quantities from point cloud based on a linear fitting of normal vectors. *Science in China Series F: Information Sciences*, *52*(3), 431–444.

29. Chiu, T. W., & Hsieh, T. H., (2006). TWQCD collaboration. Pseudovector meson with strangeness and closed charm. *Physical Review D*, *73*(11), 111503.

30. Chou, K. C., (2009). On the pseudovector current and lepton decays of baryons and mesons. In: *Selected Papers of KC Chou* (Vol. 1, pp. 106–111).

31. Clerk-Maxwell, J., (1869). Remarks on the mathematical classification of physical quantities. *Proceedings of the London Mathematical Society*, *1*(1), 224–233.

32. Cline, M. J., (1985). Perspectives for gene therapy: Inserting new genetic information into mammalian cells by physical techniques and viral vectors. *Pharmacology and Therapeutics*, *29*(1), 69–92.

33. Close, F. E., & Törnqvist, N. A., (2002). Scalar mesons above and below 1 GeV. *Journal of Physics G: Nuclear and Particle Physics*, *28*(10), R249.

34. Costello, E. K., Lauber, C. L., Hamady, M., Fierer, N., Gordon, J. I., & Knight, R., (2009). Bacterial community variation in human body habitats across space and time. *Science*, *326*(5960), 1694–1697.

35. Coutinho, M. M., Hoskins, B. J., & Buizza, R., (2004). The influence of physical processes on extratropical singular vectors. *Journal of the Atmospheric Sciences*, *61*(2), 195–209.

36. Couture, G., & König, H., (1996). Bounds on second generation scalar leptoquarks from the anomalous magnetic moment of the muon. *Physical Review D*, *53*(1), 555.

37. Cremers, D., & Soatto, S., (2003). Variational space-time motion segmentation. In: *Null* (Vol. 1, pp. 886). IEEE.

38. Cremmer, E., Julia, B., Scherk, J., Van Nieuwenhuizen, P., Ferrara, S., & Girardello, L., (1978). Super-Higgs effect in supergravity with

general scalar interactions. *Physics Letters B*, *79*(3), 231–234.

39. Croyle, M. A., Cheng, X., & Wilson, J. M., (2001). Development of formulations that enhance physical stability of viral vectors for gene therapy. *Gene Therapy*, *8*(17), 1281–1290.

40. Da Rold, L., & Pomarol, A., (2006). The scalar and pseudoscalar sector in a five-dimensional approach to chiral symmetry breaking. *Journal of High Energy Physics*, *2006*(01), 157.

41. Deaton, J. E., Barba, C., Santarelli, T., Rosenzweig, L., Souders, V., McCollum, C., & Singer, M. J., (2005). Virtual environment cultural training for operational readiness (VECTOR). *Virtual Reality*, *8*(3), 156–167.

42. Debunne, G., Desbrun, M., Cani, M. P., & Barr, A. H., (2001). Dynamic real-time deformations using space and time adaptive sampling. In: *Proceedings of the 28th Annual Conference on Computer Graphics and Interactive Techniques* (Vol. 1, pp. 31–36). ACM.

43. Deo, B. B., & Bisoi, A. K., (1974). Pseudoscalar-versus-pseudovector interactions in photo-and electroproduction of charged mesons. *Physical Review D*, *9*(1), 288.

44. Diebel, J., (2006). Representing attitude: Euler angles, unit quaternions, and rotation vectors. *Matrix*, *58*(15/16), 1–35.

45. Divotgey, F., Olbrich, L., & Giacosa, F., (2013). Phenomenology of axial-vector and pseudovector mesons: Decays and mixing in the kaonic sector. *The European Physical Journal A*, *49*(10), 135.

46. Döring, A., & Isham, C. J., (2008). A topos foundation for theories of physics: III. The representation δ □ o (A): Σ □→ R□ □. *Journal of Mathematical Physics*, *49*(5), 053517.

47. Dupree, R., Ford, N., & Holland, D., (1987). An examination of the 29 Si environment in the PbO-SiO$_2$ system by magic angle spinning nuclear magnetic resonance. *Physics and Chemistry of Glasses*, *28*(2), 78–84.

48. Durrer, R., Gasperini, M., Sakellariadou, M., & Veneziano, G., (1998). Massless (pseudo-) scalar seeds of CMB anisotropy. *Physics Letters B*, *436*(1/2), 66–72.

49. Eicher, J. B., (1995). *Dress and Ethnicity: Change Across Space and Time* (Vol. 27, pp. 1–30). Oxford: Berg.

50. Einstein, A., Podolsky, B., & Rosen, N., (1935). Can quantum-mechanical description of physical reality be considered complete?

Physical Review, *47*(10), 777.

51. Elith, J., & Leathwick, J. R., (2009). Species distribution models: Ecological explanation and prediction across space and time. *Annual Review of Ecology, Evolution, and Systematics*, *40*, 677–697.

52. Faraei, Z., & Jafari, S. A., (2017). Superconducting proximity in three-dimensional Dirac materials: Odd-frequency, pseudoscalar, pseudovector, and tensor-valued superconducting orders. *Physical Review B*, *96*(13), 134516.

53. Foote, G. B., & Du Toit, P. S., (1969). Terminal velocity of raindrops aloft. *Journal of Applied Meteorology*, *8*(2), 249–253.

54. Fujiwara, H., & Shimono, T., (1983). On the acceleration of test generation algorithms. *IEEE Transactions on Computers*, *12*, 1137–1144.

55. Fushchich, W. I., Krivsky, I. Y., & Simulik, V. M., (1989). On vector and pseudovector Lagrangians for electromagnetic field. *Il Nuovo Cimento B (1971–1996)*, *103*(4), 423–429.

56. Galton, A., (1993). Towards an integrated logic of space, time, and motion. In: *IJCAI* (Vol. 93, pp. 1550–1555).

57. Gamba, A., (1967). Physical quantities in different reference systems according to relativity. *American Journal of Physics*, *35*(2), 83–89.

58. Ge, M., Du, R., Zhang, G., & Xu, Y., (2004). Fault diagnosis using support vector machine with an application in sheet metal stamping operations. *Mechanical Systems and Signal Processing*, *18*(1), 143–159.

59. Gessler, J., (1973). Vectors in dimensional analysis. *Journal of the Engineering Mechanics Division*, *99*(1), 121–129.

60. Ghorbani, K., (2015). Fermionic dark matter with pseudo-scalar Yukawa interaction. *Journal of Cosmology and Astroparticle Physics*, *2015*(01), 015.

61. Glück, M., (1972). Pseudovector coupling and charged-pion electroproduction. *Physical Review Letters*, *28*(22), 1486.

62. Gnecco, E., Bennewitz, R., Gyalog, T., Loppacher, C., Bammerlin, M., Meyer, E., & Güntherodt, H. J., (2000). Velocity dependence of atomic friction. *Physical Review Letters*, *84*(6), 1172.

63. Goethals, B., Laur, S., Lipmaa, H., & Mielikäinen, T., (2004). On private scalar product computation for privacy-preserving data mining. In: *International Conference on Information Security and Cryptology*

(Vol. 1, pp. 104–120). Springer, Berlin, Heidelberg.

64. Gopalswamy, N., Lara, A., Lepping, R. P., Kaiser, M. L., Berdichevsky, D., & St. Cyr, O. C., (2000). Interplanetary acceleration of coronal mass ejections. *Geophysical Research Letters*, *27*(2), 145–148.

65. Gorelick, L., Blank, M., Shechtman, E., Irani, M., & Basri, R., (2007). Actions as space-time shapes. *IEEE Transactions on Pattern Analysis and Machine Intelligence*, *29*(12), 2247–2253.

66. Gray, R. W., & Rolland, J. P., (2015). Wavefront aberration function in terms of RV Shack's vector product and Zernike polynomial vectors. *JOSA A*, *32*(10), 1836–1847.

67. Gutenberg, B., & Richter, C. F., (1942). Earthquake magnitude, intensity, energy, and acceleration. *Bulletin of the Seismological Society of America*, *32*(3), 163–191.

68. Harichandran, R. S., & Vanmarcke, E. H., (1986). Stochastic variation of earthquake ground motion in space and time. *Journal of Engineering Mechanics*, *112*(2), 154–174.

69. Harlander, R. V., & Kilgore, W. B., (2002). Production of a pseudo-scalar Higgs boson at hadron colliders at next-to-next-to-leading order. *Journal of High Energy Physics*, *2002*(10), 017.

70. Harvey, D., (1990). Between space and time: Reflections on the geographical imagination. *Annals of the Association of American Geographers*, *80*(3), 418–434.

71. Hassibi, B., & Hochwald, B. M., (2002). High-rate codes that are linear in space and time. *IEEE Transactions on Information Theory*, *48*(7), 1804–1824.

72. Hauser, W., (1986). Vector products and pseudovectors. *American Journal of Physics*, *54*(2), 168–172.

73. He, H. J., & Yuan, C. P., (1999). New method for detecting charged scalars at colliders. *Physical Review Letters*, *83*(1), 28.

74. Hecking, P., Brockmann, R., & Weise, W., (1978). Pseudoscalar and pseudovector coupling of a pion to a bound Dirac nucleon. *Physics Letters B*, *72*(4), 432–435.

75. Hodgkin, A. L., (1954). A note on conduction velocity. *The Journal of Physiology*, *125*(1), 221–224.

76. Hong, D. K., & Kim, D., (2009). Pseudo scalar contributions to light-by-light correction of muon g-2 in AdS/QCD. *Physics Letters B*, *680*(5), 480–484.

77. Hsiao, S. S., Lu, D. H., & Yang, S. N., (2000). Pseudovector versus pseudoscalar coupling in kaon photoproduction reexamined. *Physical Review C*, *61*(6), 068201.

78. Hursh, J. B., (1939). Conduction velocity and diameter of nerve fibers. *American Journal of Physiology-Legacy Content*, *127*(1), 131–139.

79. Inoue, K., Kakuto, A., & Takano, H., (1986). Higgs as (pseudo-) Goldstone particles. *Progress of Theoretical Physics*, *75*(3), 664–676.

80. International Bureau of Weights and Measures, Taylor, B. N., & Thompson, A., (2001). *The International System of Units (SI)* (Vol. 1, pp. 1–50). US Department of Commerce, Technology Administration, National Institute of Standards and Technology.

81. Iwamoto, N., (1989). Pseudoscalar versus pseudovector pion-nucleon couplings in the nucleonic axion bremsstrahlung rate of neutron-star matter. *Physical Review D*, *39*(8), 2120.

82. Jetz, W., Thomas, G. H., Joy, J. B., Hartmann, K., & Mooers, A. O., (2012). The global diversity of birds in space and time. *Nature*, *491*(7424), 444.

83. Johansson, I., (2009). Mathematical vectors and physical vectors. *Dialectica*, *63*(4), 433–447.

84. Jones, F. C., & Ellison, D. C., (1991). The plasma physics of shock acceleration. *Space Science Reviews*, *58*(1), 259–346.

85. Kamiński, R., Leśniak, L., & Rybicki, K., (1997). Separation of S-wave pseudoscalar and pseudovector amplitudes in $\pi-$ P↑→ $\pi+$ $\pi-$ n reaction on polarized target. *Zeitschrift für Physik C Particles and Fields*, *74*(1), 79–91.

86. Karasuyama, H., & Melchers, F., (1988). Establishment of mouse cell lines which constitutively secrete large quantities of interleukin 2, 3, 4, or 5, using modified cDNA expression vectors. *European Journal of Immunology*, *18*(1), 97–104.

87. Kelly, P. F., Carrington, J., Nathwani, A., & Vanin, E. F., (2001). RD114-pseudotyped oncoretroviral vectors: Biological and physical properties. *Annals of the New York Academy of Sciences*, *938*(1), 262–277.

88. Kholodenko, B. N., Hancock, J. F., & Kolch, W., (2010). Signaling ballet in space and time. *Nature Reviews Molecular Cell Biology*, *11*(6), 414.

89. Knyazev, A. V., & Argentati, M. E., (2002). Principal angles between

subspaces in an A-based scalar product: Algorithms and perturbation estimates. *SIAM Journal on Scientific Computing*, *23*(6), 2008–2040.

90. Kraichnan, R. H., (1970). Diffusion by a random velocity field. *The Physics of Fluids*, *13*(1), 22–31.

91. Kruger, J., & Westermann, R., (2003). Acceleration techniques for GPU-based volume rendering. In: *Proceedings of the 14th IEEE Visualization 2003 (VIS'03)* (Vol. 14, pp. 38). IEEE Computer Society.

92. Lahanas, A. B., & Spanos, V. C., (2002). Implications of the pseudo-scalar Higgs boson in determining the neutralino dark matter. *The European Physical Journal C-Particles and Fields*, *23*(1), 185–190.

93. Laptev, I., (2005). On space-time interest points. *International Journal of Computer Vision*, *64*(2/3), 107–123.

94. Leurer, M., (1994). Bounds on vector leptoquarks. *Physical Review D*, *50*(1), 536.

95. Li, X., & Hodgson, M. E., (2004). Vector field data model and operations. *GI Science and Remote Sensing*, *41*(1), 1–24.

96. Li, Y., Liou, M. K., & Schreiber, W. M., (1998). Proton-proton bremsstrahlung calculation: Studies of the off-shell proton electromagnetic vertex and of pseudoscalar vs pseudovector π N couplings. *Physical Review C*, *57*(2), 507.

97. Liehr, S., Muanenda, Y. S., Münzenberger, S., & Krebber, K., (2017). Relative change measurement of physical quantities using dual-wavelength coherent OTDR. *Optics Express*, *25*(2), 720–729.

98. Lighthill, M. J., (1970). Group velocity. In: *Hyperbolic Equations and Waves* (Vol. 1, pp. 96–123). Springer, Berlin, Heidelberg.

99. Lipkin, H., & Duffy, J. O. S. E. P. H., (1982). Analysis of industrial robots via the theory of screws. *Proc. 12th Int. Sym. Indus. Robots*, *12*(1), 359–370.

100. Loarie, S. R., Duffy, P. B., Hamilton, H., Asner, G. P., Field, C. B., & Ackerly, D. D., (2009). The velocity of climate change. *Nature*, *462*(7276), 1052.

101. Luh, J., Walker, M., & Paul, R., (1980). Resolved-acceleration control of mechanical manipulators. *IEEE Transactions on Automatic Control*, *25*(3), 468–474.

102. MacDonald, D., & Thorne, K. S., (1982). Black-hole electrodynamics: An absolute-space/universal-time formulation. *Monthly Notices of the Royal Astronomical Society*, *198*(2), 345–382.

103. Mackey, D. S., Mackey, N., & Tisseur, F., (2004). G-reflectors: Analogues of householder transformations in scalar product spaces. *Linear Algebra and its Applications*, *385*, 187–213.

104. Mallik, R. K., (2008). Distribution of inner product of two complex Gaussian vectors and its application to MPSK performance. In: *2008 IEEE International Conference on Communications* (Vol.1, pp. 4616–4620). IEEE.

105. Mäntylä, T., & Koponen, I. T., (2007). Understanding the role of measurements in creating physical quantities: A case study of learning to quantify temperature in physics teacher education. *Science and Education*, *16*(3–5), 291–311.

106. Marghitu, D. B., (2005). *Kinematic Chains and Machine Components Design* (Vol. 11, pp. 9–21). Gulf Professional Publishing.

107. Maris, P., & Roberts, C. D., (1998). Pseudovector components of the pion, π 0$\rightarrow$ γ γ, and F π (q 2). *Physical Review C*, *58*(6), 3659.

108. Maznevski, M. L., & Chudoba, K. M., (2000). Bridging space over time: Global virtual team dynamics and effectiveness. *Organization Science*, *11*(5), 473–492.

109. McConnell, H. M., (1958). A pseudovector nuclear hyperfine interaction. *Proceedings of the National Academy of Sciences of the United States of America*, *44*(8), 766.

110. McDicken, W. N., Sutherland, G. R., Moran, C. M., & Gordon, L. N., (1992). Color Doppler velocity imaging of the myocardium. *Ultrasound in Medicine and Biology*, *18*(6–7), 651–654.

111. Mendiratta, M. G., & Dimiduk, D. M., (1991). Phase relations and transformation kinetics in the high Nb region of the Nb-Si system. *Scripta Metallurgica*, *25*(1), 237–242.

112. Miller, R. G., & Phillips, R. A., (1969). Separation of cells by velocity sedimentation. *Journal of Cellular Physiology*, *73*(3), 191–201.

113. Muller, P., (1998). Space-time as a primitive for space and motion. In *FOIS'98* (Vol. 1, pp. 63–76).

114. Musacchio, A., & Salmon, E. D., (2007). The spindle-assembly checkpoint in space and time. *Nature Reviews Molecular Cell Biology*, *8*(5), 379.

115. Nespor, J., (2014). *Knowledge in Motion: Space, Time and Curriculum in Undergraduate Physics and Management* (Vol. 2, pp. 1–20). Routledge.

116. Niu, H., Matsuda, T., Sadamatsu, H., & Takai, M., (1976). Application of lateral photovoltaic effect to the measurement of the physical quantities of PN junctions–sheet resistivity and junction conductance of N_2^+ implanted Si. *Japanese Journal of Applied Physics, 15*(4), 601.

117. Noble, J. V., (1979). Pseudoscalar-pseudovector equivalence of threshold production and absorption of pions. *Physical Review Letters, 43*(2), 100.

118. Norton, R. E., & Watson, W. K. R., (1958). Consequences of a pseudovector pion-nucleon coupling and the universal β decay. *Physical Review, 110*(4), 996.

119. Noui, K., & Perez, A., (2005). Three-dimensional loop quantum gravity: Physical scalar product and spin-foam models. *Classical and Quantum Gravity, 22*(9), 1739.

120. Page, C. H., & Vigoureux, P., (1977). *The International System of Units (SI)* (Vol. 330, pp. 1–40). US Department of Commerce, National Bureau of Standards.

121. Palmer, D. J., & Ng, P., (2004). Physical and infectious titers of helper-dependent adenoviral vectors: A method of direct comparison to the adenovirus reference material. *Molecular Therapy, 10*(4), 792–798.

122. Pavlačka, O., & Talašová, J., (2010). Fuzzy vectors as a tool for modeling uncertain multidimensional quantities. *Fuzzy Sets and Systems, 161*(11), 1585–1603.

123. Pearman, P. B., Guisan, A., Broennimann, O., & Randin, C. F., (2008). Niche dynamics in space and time. *Trends in Ecology and Evolution, 23*(3), 149–158.

124. Perrine, J. J., & Edgerton, V. R., (1978). Muscle force-velocity and power-velocity relationships under isokinetic loading. *Medicine and Science in Sports, 10*(3), 159–166.

125. Philippsen, P., Kramer, R. A., & Davis, R. W., (1978). Cloning of the yeast ribosomal DNA repeat unit in SstI and HindIII lambda vectors using genetic and physical size selections. *Journal of Molecular Biology, 123*(3), 371–386.

126. Pickett, S. T., (1989). Space-for-time substitution as an alternative to long-term studies. In: *Long-Term Studies in Ecology* (Vol. 1, pp. 110–135). Springer, New York, NY.

127. Poghossian, A., Schultze, J. W., & Schöning, M. J., (2003). Multi-parameter detection of (bio-) chemical and physical quantities using

an identical transducer principle. *Sensors and Actuators B: Chemical*, *91*(1–3), 83–91.

128. Polyak, B. T., & Juditsky, A. B., (1992). Acceleration of stochastic approximation by averaging. *SIAM Journal on Control and Optimization*, *30*(4), 838–855.

129. Pukhov, A., Sheng, Z. M., & Meyer-ter-Vehn, J., (1999). Particle acceleration in relativistic laser channels. *Physics of Plasmas*, *6*(7), 2847–2854.

130. Pulay, P., (1982). Improved SCF convergence acceleration. *Journal of Computational Chemistry*, *3*(4), 556–560.

131. Reames, D. V., (1999). Particle acceleration at the Sun and in the heliosphere. *Space Science Reviews*, *90*(3/4), 413–491.

132. Reschel, T., Koňák, Č., Oupický, D., Seymour, L. W., & Ulbrich, K., (2002). Physical properties and in vitro transfection efficiency of gene delivery vectors based on complexes of DNA with synthetic polycations. *Journal of Controlled Release*, *81*(1/2), 201–217.

133. Richert, R., (2002). Heterogeneous dynamics in liquids: Fluctuations in space and time. *Journal of Physics: Condensed Matter*, *14*(23), R703.

134. Richtmyer, R. D., (1960). Taylor instability in shock acceleration of compressible fluids. *Communications on Pure and Applied Mathematics*, *13*(2), 297–319.

135. Rideout, D. C., & Breslow, R., (1980). Hydrophobic acceleration of Diels-Alder reactions. *Journal of the American Chemical Society*, *102*(26), 7816–7817.

136. Rose, C., Guenter, B., Bodenheimer, B., & Cohen, M. F., (1996). Efficient generation of motion transitions using space-time constraints. In: *Siggraph* (Vol. 96, pp. 147–154).

137. Rossow, W. B., & Zhang, Y. C., (1995). Calculation of surface and top of atmosphere radiative fluxes from physical quantities based on ISCCP data sets: 2. Validation and first results. *Journal of Geophysical Research: Atmospheres*, *100*(D1), 1167–1197.

138. Rude, M., (1997). Collision avoidance by using space-time representations of motion processes. *Autonomous Robots*, *4*(1), 101–119.

139. Rynasiewicz, R., (1995). By their properties, causes and effects: Newton's scholium on time, space, place and motion—I. The text. *Studies in History and Philosophy of Science Part A*, *26*(1), 133–153.

140. Salmon, P., Arrighi, J. F., Piguet, V., Chapuis, B., Zubler, R. H., Trono, D., & Kindler, V., (2001). Transduction of CD34+ cells with lentiviral vectors enables the production of large quantities of transgene-expressing immature and mature dendritic cells. *The Journal of Gene Medicine: A Cross-Disciplinary Journal for Research on the Science of Gene Transfer and its Clinical Applications*, *3*(4), 311–320.

141. Salomone, A., Schechter, J., & Tudron, T., (1981). Properties of scalar gluonium. *Physical Review D*, *23*(5), 1143.

142. Sazdjian, H., (1988). The scalar product in two-particle relativistic quantum mechanics. *Journal of Mathematical Physics*, *29*(7), 1620–1633.

143. Scott, J. F., (1971). Phonon-plasmon decoupling: Bound exciton resonant Raman effect in CdS. *Solid State Communications*, *9*(11), 759–764.

144. Sexton, J., Vaccarino, A., & Weingarten, D., (1995). Numerical evidence for the observation of a scalar glueball. *Physical Review Letters*, *75*(25), 4563.

145. Shechtman, E., & Irani, M., (2007). Space-time behavior-based correlation-or-how to tell if two underlying motion fields are similar without computing them? *IEEE Transactions on Pattern Analysis and Machine Intelligence*, *29*(11), 2045–2056.

146. Silva, C. C., & De Andrade, M. R., (2002). Polar and axial vectors versus quaternions. *American Journal of Physics*, *70*(9), 958–963.

147. Snyder, H. S., (1947). Quantized space-time. *Physical Review*, *71*(1), 38.

148. Sprangle, P., Joyce, G., Esarey, E., & Ting, A., (1988). Laser wake field acceleration and relativistic optical guiding. In: *AIP Conference Proceedings* (Vol. 175, No. 1, pp. 231–239). AIP.

149. Spurr, R., (2008). LIDORT and VLIDORT: Linearized pseudo-spherical scalar and vector discrete ordinate radiative transfer models for use in remote sensing retrieval problems. In: *Light Scattering Reviews 3* (pp. 229–275). Springer, Berlin, Heidelberg.

150. Stampfli, G., Marcoux, J., & Baud, A., (1991). Tethyan margins in space and time. *Paleogeography, Palaeoclimatology, Palaeoecology*, *87*(1–4), 373–409.

151. Steffen, W., Broadgate, W., Deutsch, L., Gaffney, O., & Ludwig, C., (2015). The trajectory of the Anthropocene: the great acceleration. *The*

Anthropocene Review, *2*(1), 81–98.

152. Stetson, K. A., (1975). Fringe vectors and observed-fringe vectors in hologram interferometry. *Applied Optics*, *14*(2), 272–273.

153. Sugawara, M., & Okubo, S., (1960). Two-nucleon potential from pion field theory with pseudovector coupling. *Physical Review*, *117*(2), 611.

154. Swanton, C., (2012). Intratumor heterogeneity: Evolution through space and time. *Cancer Research*, *72*(19), 4875–4882.

155. Tagliani, A., (2003). Maximum entropy solutions and moment problem in unbounded domains. *Applied Mathematics Letters*, *16*(4), 519–524.

156. Taner, M. T., & Koehler, F., (1969). Velocity spectra—digital computer derivation applications of velocity functions. *Geophysics*, *34*(6), 859–881.

157. Tanner, J. M., & Whitehouse, R. H., (1976). Clinical longitudinal standards for height, weight, height velocity, weight velocity, and stages of puberty. *Archives of Disease in Childhood*, *51*(3), 170–179.

158. Taylor, B. N., & Mohr, P. J., (2001). The role of fundamental constants in the international system of units (SI): Present and future. *IEEE Transactions on Instrumentation and Measurement*, *50*(2), 563–567.

159. Thrift, N. J., (1983). On the determination of social action in space and time. *Environment and Planning D: Society and Space*, *1*(1), 23–57.

160. Toulmin, S., (1959). Criticism in the history of science: Newton on absolute space, time, and motion, I. *The Philosophical Review*, *68*(1), 1–29.

161. Truesdell, C., (1953). The physical components of vectors and tensors. *ZAMM-Journal of Applied Mathematics and Mechanics/Zeitschrift für Angewandte Mathematik und Mechanik*, *33*(10/11), 345–356.

162. Tsiatas, G. C., (2009). A new Kirchhoff plate model based on a modified couple stress theory. *International Journal of Solids and Structures*, *46*(13), 2757–2764.

163. Valassi, A., (2003). Combining correlated measurements of several different physical quantities. *Nuclear Instruments and Methods in Physics Research Section A: Accelerators, Spectrometers, Detectors and Associated Equipment*, *500*(1–3), 391–405.

164. Van Assendelft, O. W., Mook, G. A., & Zijlstra, W. G., (1973). International system of units (SI) in physiology. *Pflügers Archiv*, *339*(4), 265–272.

165. Van Beveren, E., Rijken, T. A., Metzger, K., Dullemond, C., Rupp, G., & Ribeiro, J. E., (1986). A low-lying scalar meson nonet in a unitarized meson model. *Zeitschrift für Physik C Particles and Fields, 30*(4), 615–620.

166. Van der Vegt, J. J., & Van der Ven, H., (2002). Space-time discontinuous Galerkin finite element method with dynamic grid motion for inviscid compressible flows: I. General formulation. *Journal of Computational Physics, 182*(2), 546–585.

167. Van Hove, L., (1954). Correlations in space and time and born approximation scattering in systems of interacting particles. *Physical Review, 95*(1), 249.

168. Van Loo, F. J., Smet, F. M., Rieck, G. D., & Verspui, G., (1982). Phase relations and diffusion paths in the Mo-Si-C system at 12000 C. *High Temperatures-High Pressures, 14*(1), 25–31.

169. Vasco, D. W., (1990). Moment-tensor invariants: Searching for non-double-couple earthquakes. *Bulletin of the Seismological Society of America, 80*(2), 354–371.

170. Wakker, B. P., & van Woerden, H., (1997). High-velocity clouds. *Annual Review of Astronomy and Astrophysics, 35*(1), 217–266.

171. Weinberg, D. H., Mortonson, M. J., Eisenstein, D. J., Hirata, C., Riess, A. G., & Rozo, E., (2013). Observational probes of cosmic acceleration. *Physics Reports, 530*(2), 87–255.

172. Weiskopf, D., Hopf, M., & Ertl, T., (2001). Hardware-accelerated visualization of time-varying 2D and 3D vector fields by texture advection via programmable per-pixel operations. In: *Proceedings of the Vision Modeling and Visualization Conference 2001* (Vol.1, pp. 439–446). Aka GmbH.

173. Whitney, H., (1968). The mathematics of physical quantities: Part I: Mathematical models for measurement. *The American Mathematical Monthly, 75*(2), 115–138.

174. Widnall, S., (2009). Lecture 13-vectors, matrices and coordinate transformations. *Dynamics, 1*(1), 1–15.

175. Williams, M. Z., & Stein, F. M., (1964). A triple product of vectors in four-space. *Mathematics Magazine, 37*(4), 230–235.

176. Winans, J. G., (1977). Quaternion physical quantities. *Foundations of Physics, 7*(5/6), 341–349.

177. Woodard, S. E., & Taylor, B. D., (2007). Measurement of multiple unrelated physical quantities using a single magnetic field response sensor. *Measurement Science and Technology*, *18*(5), 1603.

178. Yang, Z., Wright, R. N., & Subramaniam, H., (2006). Experimental analysis of a privacy-preserving scalar product protocol. *International Journal of Computer Systems Science and Engineering*, *21*(1), 47–52.

179. Yoccoz, N. G., Nichols, J. D., & Boulinier, T., (2001). Monitoring of biological diversity in space and time. *Trends in Ecology and Evolution*, *16*(8), 446–453.

180. Zhang, Y. C., Rossow, W. B., & Lacis, A. A., (1995). Calculation of surface and top of atmosphere radiative fluxes from physical quantities based on ISCCP data sets: 1. Method and sensitivity to input data uncertainties. *Journal of Geophysical Research: Atmospheres*, *100*(D1), 1149–1165.

Chapter

4

Fundamentals of Engineering Mechanics

CONTENTS

4.1. INTRODUCTION

Mechanics is the particular area of science that is associated with behavior of the physical bodies when exposed to displacements or forces, and the subsequent effects of bodies on their surroundings (Inman and Singh, 1994; McCormick, 2009; Meriam and Kraige, 2012). This area of science has its backgrounds in Ancient Greece with writings of Archimedes and Aristotle. During the primary modern period, scientists like Galileo, Newton, and Kepler laid the basis for what is now acknowledged as the classical mechanics (Goodman, 1989; Daniel et al., 1994; Kutz, 2015). It is a branch of the classical physics that tackles with the particles that are either moving with velocities considerably less as compared to the speed of light or at rest. It can be defined as the branch of science that accounts for the motion and forces on the objects (Eshbach and Tapley, 1990; Bedford et al., 2003; Hibbeler and Fan, 2004). The field is still less extensively understood in terms of quantum theory (Bardet, 1998; Ma and Zhang, 2016). Essential concepts and quantities of the mechanics are explained below (Streeter Jr and Hanna, 1973; Muhanna et al., 2007):

4.1.1. Space

Three dimensional geometrical regions, where the objects are placed and move. The measure of size of the physical system and of distances amongst objects is known as the length. The position of the point in the space can be found comparative to a reference point by utilizing linear and measurements with respect to the coordinate system whose starting point is at the reference point (Vable, 1985; Muhanna and Mullen, 2001; Boresi et al., 2010).

4.1.2. Time

The abstract concept utilized to describe the simultaneity of the events. The measure of the interval amongst two events.

4.1.3. Matter

Matter is the any substance that covers space. A body is a matter bounded by closed surface.

4.1.4. Mass

Mass is the measure of quantities of the matter. Mass is the measure of inertia described in its turn as the body's resistance to the change in motion.

4.1.5. Force

The abstract concept used to measure the action amongst two bodies consequential in change in the motions.

4.1.6. Particle

It is an object without size or shape but with mass. The model of real body moving in the distances much larger than the size of body.

4.1.7. Rigid Body

It is a collection of the particles, whose comparative distances are constant (Haber, 1990; Jian-Le, 2009).

4.2. MECHANICAL MODELS

Mechanics can be described as the *science of the motion of the bodies*. Instead of utilizing real objects, mechanics utilize their *models*. Generally, the model of given object is an image reflecting those characteristics of the object which are important to inspect the phenomena of interest for specific branch of science (Benioff, 1982; Lim et al., 2006). To basic models applied in the mechanics belong the subsequent ones:

4.2.1. A Particle

A particle is a body having mass but having very small dimensions that the body can be preserved as a point in the geometric sense. However, in exercise, bodies having zero angular velocities by assumption or the bodies whose rotational motion can be ignored are treated as the particles irrespective of their dimensions (Phillips, 1950; Jones, 1989).

4.2.2. System of Particles

System of particles is a collection of the particles. The system can be composed of a collection of finite particles.

4.2.3. A Rigid Body

The distances among elements of such kind of a body remain constant for randomly large magnitudes of the forces acting on body. Machines, structures, and mechanisms are deformable bodies. However, normally the deformations are small, and therefore in many circumstances their effect on dynamics/statics of studied bodies can be ignored (Hinch, 1977; Goldberger, 1989).

4.2.4. System of Rigid Bodies

It is a collection of rigid bodies. The laws of mechanics presented by Newton assist to brighten the motions of the material systems. They help us to create the *mathematical model* to formulate the *equations of motion* of the bodies and particles. The main objective of engineering mechanics is to work out the laws of motion appropriate for investigation of the diversity of real bodies. Any real body, solid, gaseous, or liquid, can be modeled as the collection of particles (Ohayon and Soize, 1997; Lomov et al., 2000). The below-mentioned branches of the mechanics deal with issues in the earlier mentioned fields:

- Mechanics of the rigid bodies (statics and dynamics).
- Mechanics of the deformable bodies (strength of the materials, plastic theory, elasticity theory, or rheology).
- Mechanics of the fluids: Incompressible and compressible; mechanics of the incompressible fluids like water is called the hydraulics.

Engineering mechanics deals with the process of modeling the geometry of mass and description of the materials from which the bodies are formed. In the rigid-body mechanics, it is assumed that the distance amongst any two points of the body doesn't change. An entirely different problem is talked about when there is a probability of varying the distance among the points of the body. The load of the bodies, in this case, leads to variation in the distance between body atoms, and the interatomic forces will stable the external load (Lister and Kerr, 1991). Bodies and the material systems made up of metal as the encountered in technology possess regular structures of the arranged atomic networks in order of 10^{30}. Regarding large amounts of the atoms, analysis is done on micro-scale, which leads to averaging of anisotropy of the microcrystal systems. Generally, most of the technical materials, after having the cubicoid cut out with sides of roughly 10^3 meters, have the similar properties regardless of the direction of the cutting out, and such materials are known as isotropic. There also exist anisotropic materials in technology whose lasting properties are dependent on the direction in which cube of the material is cut out (Lagomarsino and Giovinazzi, 2006).

4.3. CLASSICAL AND THE QUANTUM MECHANICS

Historically, classical mechanics was introduced first and quantum mechanics is a relatively recent development. Classical mechanics instigated with Isaac Newton's laws of the motion in Philosophiæ Naturalis Principia Mathematica; whereas quantum Mechanics was established in the early twentieth century (Casati et al., 1987; Maslov and Fedoriuk, 2001). Both are usually held to establish the most certain understanding that exists in physical nature.

Classical mechanics has particularly often been observed as a model for the other so-called exact sciences. Important in this respect is broad utilization of mathematics in the theories, along with the decisive part played by experiments in producing and testing them (Suppes, 1966; Beltrametti and Bugajski, 1995).

Quantum mechanics the scope is quite different, as classical mechanics could be thought as the behavior of nature under certain constrained circumstances. According to the correspondence principle, there isn't any contradiction or conflict among the two subjects, each pertains to particular situations. The correspondence principle defines that behavior of the systems described by the quantum theories replicates classical physics in limit of the large quantum numbers (Leaf, 1968; Man'ko, 1996). Quantum mechanics has outdated classical mechanics at the level of foundation and is necessary for explanation and forecast of the processes at atomic, molecular, and sub-atomic level. However, for the macroscopic procedures classical mechanics is capable to solve issues that are waywardly difficult in quantum mechanics and therefore remains helpful and well used. Modern descriptions of such behavior initiate with the careful definition of quantities as displacement, time, mass, velocity, acceleration, and force. Until some 400 years ago, motion was described from a very different viewpoint, for instance, following the thoughts of Greek philosopher and the scientist Aristotle, scientists articulated that the cannonball falls down as its position is in the Earth; sun, moon, and the stars move in circle around the earth since it is the nature of these objects to move in perfect circles (Hepp, 1974; Habib, 1990).

Frequently cited as a father to the modern science, Galileo merged the thoughts of the great intellectuals of his time and started to calculate the motion in terms of the distance change from any starting position and time. He stated that the speed of the falling objects increases gradually during the time of the fall. This acceleration is same for the heavy objects as for the light ones, provided air friction is not considered. The English mathematician and the physicist Lord Isaac Newton enhanced this analysis by describing mass and force and linking these to the acceleration. For objects moving at the speeds near to speed of the light, Newton's laws of motion were succeeded by Albert Einstein's theory of the relativity. For atomic and the subatomic particles, Newton's laws of motion were succeeded by quantum theory. For everyday phenomena, though, Newton's three laws of motion stay the cornerstone of the dynamics, which is the examination of what causes the motion (Strocchi, 1966; Robert, 1997).

4.4. PRINCIPLES OF THE MECHANICS (NEWTON'S LAWS – 1687)

Following are the common principles of the mechanics founded on Newton laws (Scheck, 2010; Cannon and Dostrovsky, 2012).

- In absence of the external forces, particle initially at rest or moving with the constant velocity will stay at rest or will move with the constant velocity along the straight line (Baigrie, 1988).
- If an external force is applied on the particle, the particle will usually be accelerated in orientation of the force and magnitude of the acceleration will generally be proportional to force and inversely proportional to mass of the particle (Grattan-Guinness, 1988).
- For every action, there exists an equal and the opposite reaction. The forces of the action and reaction amongst the contacting bodies are equal in the magnitude, opposite in the direction and collinear (Gauld, 1993).
- Different forces acting on the particle at the same time act independently on one another and their outcome is also independent. Two forces can be substituted by the equivalent force determined by the parallelogram (Guicciardini, 2005).
- Law of the Gravitation: two particles having mass m_1 and m_2 at distance r attracts jointly with force proportional to the $m_1 m_2$ (product) and inversely proportional to square of the distance.

4.5. VECTORS

In mechanics like all physical sciences and the engineering fields – any experiments, research, and study are executed expressing properties of the objects and procedures with numbers (Nelson, 1959; Johansen, 1988; Kiros et al., 2015). In such kind of modeling, physical quantities are used, which are the abstract terms selected during the development of science. A physical quantity should have its experimental appearances which assist to compare different phenomena and objects in given aspect (An et al., 1989; Karimi et al., 2002). All mechanical phenomena and states can be described and defined using two types of quantities: scalars and vectors (Miller, 1992; Pennington et al., 2014).

4.5.1. Physical and Geometrical Explanation of the Vectors

- Characteristics of the vector (Bevan, 1984):
- Magnitude;
- Slope;
- Sense;
- Point of the application.
- Types of the vectors (Graham and Prevec, 1991):
- **Free Vectors:** Whose point of the application can be located anywhere and the line of action can also be moved in parallel in space;
- **Sliding Vectors:** Whose point of the application can be located somewhere on its line of the action;
- **Fixed Vector:** Have a defined point of the application and the line of action.

4.5.2. Analytical Interpretation of the Vector Quantity

In mathematics, vectors are considered as free vectors (Roe, 1981). For the analytical operations on occasion, it is easy to break into two factors concerning: direction and magnitude (sense, slope):

$$\mathbf{a} = a \cdot \mathbf{a}^o \qquad (1)$$

a – Magnitude of vector.

$\mathbf{a}^o$ – Unity vector of same sense and slope as the $\mathbf{a}$ ($\mathbf{a}$ normalized to one)

Algebraically vector can be well-defined as a triplet of the numbers whose meaning is dependent on the coordinates practical in the given case. In the Cartesian, coordinate system:

$$\mathbf{a} = \begin{vmatrix} a_x & a_y & a_z \end{vmatrix},$$

(2)

where a_x, a_y, a_z are the orthogonal projections of the vector $\mathbf{a}$

Another form can be made as the sum of 3 orthogonal vectors—the rectangular components:

$$\mathbf{a} = \mathbf{a}_x + \mathbf{a}_y + \mathbf{a}_z$$

(3)

The above equation can also be expressed utilizing scalar components and the unity vectors of the axes: $\mathbf{i}\ \mathbf{j}\ \mathbf{k}$

$$\mathbf{a} = \mathbf{i}a_x + \mathbf{j}a_y + \mathbf{k}a_z$$

(4)

Another beneficial form is:

$$\mathbf{a} = a\begin{bmatrix} \cos\acute{a} & \cos\hat{a} & \cos\tilde{a} \end{bmatrix}$$

(5)

where the unity vector of same sense and the slope $\mathbf{a}^o$ is articulated with the help of cosines of the direction of the vector's line of action:

$$\mathbf{a}^o = \mathbf{i}\cos\alpha + \mathbf{j}\cos\beta + \mathbf{k}\cos\gamma$$

(6)

$$\mathbf{a}^o = \begin{bmatrix} \cos\acute{a} & \cos\hat{a} & \cos\tilde{a} \end{bmatrix}$$

(7)

4.5.3. Specifying and Expressing the Vector

Vectors might be expressed:

- Either in the geometry with the help of parameters of forms (4.1) and (4.5).
- Or in algebra with the help of parameters of the form (4.4).

In practical problems, the vectors are given straight—through values of parameters appropriate for the selected form or with an equation, which the vector should satisfy.

4.6. NEWTON LAWS

The laws formulated originally by Newton produced a set of other important laws of mechanics like the conservation of the linear momentum, conservation of the angular momentum, and conservation of the kinetic energy (Garbaczewski, 1993; Verlinde, 2011; Cari et al., 2016). Below mentioned are the laws are given by Newton, which are useable for particles.

4.6.1. First Law

The body at rest will remain at rest unless acted upon by an external force, and the body in motion moving at constant speed along the straight line will continue in that motion until an external force acts on the body (Pascale et al., 2000; McIntyre et al., 2001).

4.6.2. Second Law

The acceleration of the particle is proportional to net force acting on the particle; the sense and the direction of acceleration are equal to those of force (Watkins, 1997; Yeo and Zadnik, 2000).

4.6.3. Third Law

The joint forces of action and the reaction amongst two bodies are equal, collinear, and opposite. The first two laws are valid in an inertial system, while this law is compulsory in any system. It can be displayed that Newton's first law is the particular circumstance of his second law. It should be observed that the laws of Newton are based on the concept of *force* as the vector quantity. Force seems here as the primitive notion and needs the introduction of two bodies as a minimum. Correlation of reactions amongst body's outcomes from third law of Newton, where the action produces an instant reaction, which is characterized graphically by description presented by the Newton: "If I exert pressure with the finger upon stone with the certain force, then stone also exerts pressure on my finger with the exact same force." The interaction of the bodies can be realized by direct pressure of one body on the other or by the indirect reaction at some distance (White, 1984; Newton, 2002).

The latter case is linked with Newton's law of the gravitation since, if two particles are considered having masses m_1 and m_2, the force of gravity F_{12} by which the particle of mass m_2 draws the particle of mass m_1 is given as:

$$F_{12} = \frac{Gm_1 m_2 \boldsymbol{r}_{12}}{r^3}$$

(8)

where G is equal to 6:67 10^{11} Nm^2 kg^2, and $\boldsymbol{r}_{12}$ is the vector position between these two points and is directed from point 1 towards point 2.

The gravitational constant was discovered by Henry Cavendish and give the order of magnitude of the interaction. It should be observed, however, that is certain arbitrariness in definition of the force. Nobel laureate Richard Feynman brings attention to the point that definition of the force in the strict sense is quite difficult. This is because of the imprecise character of the Newton's second law and usually because of the imprecise character of laws of the physics.

The idea of mass can be introduced based on Newton's second law. Let's consider a random particle and smear to the particle, in turn, the forces of several magnitudes $\mathbf{F}_1$; $\mathbf{F}_2$; $\mathbf{F}_3$... $\mathbf{F}_N$. Every force produces motion of particle with the accelerations $\mathbf{a}_1$; $\mathbf{a}_2$; $\mathbf{a}_3$...$\mathbf{a}_N$, correspondingly. According to the second law of Newton, these forces are proportional to magnitudes of the forces.

$$\frac{F_1}{a_1} = \frac{F_2}{a_2} = \frac{F_3}{a_3} = \cdots = \frac{F_N}{a_N}$$

(9)

The foregoing ratios define the *inertia* of the body (particle) and describe the *mass* of the body. Recall that *weight* of the body is product of the mass of body and the acceleration of gravity. The mass described in that way is known as the *gravitational mass*. Empirical research directed by the Hungarian physicist Roland Eötvös demonstrated that *inertial* mass and the *gravitational* mass are identical. In other words, if the particle taken is positioned on Earth's surface, then the Equation (8) might be used to describe the weight **G** of particle of the mass m. That is, presenting r D R and familiarizing gD $\frac{Gm}{}$R$_2$; m$_2$ Dm, weight of particle of the mass m is **G** Dmg.

Observe that R is dependent on particle elevation and on the latitude, and therefore the value of g changes with the position of a particle. The second law of Newton can be formulated as:

ma = **F**   (10)

The third law of Newton is also called the *law of action and reaction*. It is practical for bodies in contact as well as for the bodies interacting at the distance $(\mathbf{F}_{12} = -\mathbf{F}_{21})$.

Finally, it must be observed that the three laws of Newton were presented in the modified form. Newton's original text from his work *Philosophiae Naturalis Principia Mathematica* is somewhat different. For example, Newton doesn't use the notion of the particle but that of the body. The idea of force was described by him through the series of axioms and not in the vector notation. It is worth highlighting that the idea of force was very subjective notion historically as it was linked with the distinct sensation of exertion of the muscles.

Apart from the described laws, it is reasonable to introduce various *principles* of mechanics. While the laws define relationships among mechanical quantities frequently leading to the solutions, the principles only assist the formulation of the equations of motion. The principles have the value of universality as they can be made practical, for instance, in theory of the relativity, quantum mechanics, and certain branches of the physics. They can be divided into *differential* and *integral principles*. The principles utilized in classical mechanics are part of the so-called *analytical mechanics*. The principle of the *independent force of the action* is generalization of Newton's second law. If various forces act upon the particle, the acceleration of the particle is an outcome of the geometric sum of accelerations produced by every force acting separately.

Let's recall, concluding, that the explanation of the behavior of the electromagnetic fields presented by Maxwell's equations was in deviation with Newton's concept of the particle motion. It turned out that the electromagnetic waves could travel in a vacuum. This contradicts the purely mechanical method whereby waves can travel only in the material medium filling up space. Moreover, these equations were constant with respect to Lorentz transformation, while Newton's equations are constant with respect to Galilean transformation. Albert Einstein succeeded in resolving that issue thanks to an overview of so-called special theory of the relativity in 1905. He presented space-time as a continuous and unified *quantity*, laying the basis of so-called *relativistic mechanics*. In this way, two deductive systems merged, i.e., electrodynamics, and mechanics. In the relativistic mechanics, time, space, and mass are dependent on one another and can't be treated as an *absolute independent attributes*. Fortunately, the dissimilarities amongst relativistic mechanics and Newton's mechanics seem at particle speeds near to speed of the light or in analysis of the large distances.

REFERENCES

1. An, G., Ebert, P. R., Mitra, A., & Ha, S. B., (1989). Binary vectors. In: *Plant Molecular Biology Manual* (Vol. 1, pp. 29–47). Springer, Dordrecht.

2. Baigrie, B. S., (1988). The vortex theory of motion, 1687–1713: Empirical difficulties and guiding assumptions. In: *Scrutinizing Science* (Vol. 1, pp. 85–102). Springer, Dordrecht.

3. Bardet, J. P., (1998). Introduction to computational granular mechanics. In: *Behavior of Granular Materials* (Vol. 1, pp. 99–169). Springer, Vienna.

4. Bedford, A., Fowler, W. L., & Fowler, W. T., (2003). *Engineering Mechanics: Statics and Dynamics Principles* (Vol. 1, pp. 1–20). Pearson Education.

5. Beltrametti, E. G., & Bugajski, S., (1995). A classical extension of quantum mechanics. *Journal of Physics A: Mathematical and General, 28*(12), 3329.

6. Benioff, P., (1982). Quantum mechanical models of Turing machines that dissipate no energy. *Physical Review Letters, 48*(23), 1581.

7. Bevan, M., (1984). Binary agrobacterium vectors for plant transformation. *Nucleic Acids Research, 12*(22), 8711–8721.

8. Boresi, A. P., Chong, K., & Lee, J. D., (2010). *Elasticity in Engineering Mechanics* (Vol. 1, pp. 1–16). John Wiley & Sons.

9. Cannon, J. T., & Dostrovsky, S., (2012). *The Evolution of Dynamics: Vibration Theory from 1687 to 1742: Vibration Theory from 1687 to 1742* (Vol. 6, pp. 10–56). Springer Science & Business Media.

10. Cari, C., Suparmi, A., & Handhika, J., (2016). Student's preconception and anxiety when they solve multi representation concepts in Newton laws and its application. In: *Journal of Physics: Conference Series* (Vol. 776, No. 1, p. 012091). IOP Publishing.

11. Casati, G., Chirikov, B. V., Shepelyansky, D. L., & Guarneri, I., (1987). Relevance of classical chaos in quantum mechanics: The hydrogen atom in a monochromatic field. *Physics Reports, 154*(2), 77–123.

12. Daniel, I. M., Ishai, O., Daniel, I. M., & Daniel, I., (1994). *Engineering Mechanics of Composite Materials* (Vol. 3, pp. 256–256). New York: Oxford university press.

13. Dunstan, D. J., (2008). Derivation of special relativity from Maxwell and Newton. *Philosophical Transactions of the Royal Society A: Mathematical, Physical and Engineering Sciences, 366*(1871), 1861–1865.

14. Eshbach, O. W., & Tapley, B. D., (1990). *Eshbach's Handbook of Engineering Fundamentals* (Vol. 1, pp. 1–20). John Wiley & Sons.

15. Garbaczewski, P., (1993). Physical significance of the Nelson-Newton laws. *Physics Letters A, 172*(4), 208–214.

16. Gauld, C., (1993). The historical context of Newton's third law and the teaching of mechanics. *Research in Science Education, 23*(1), 95–103.

17. Goldberger, A. S., (1989). Economic and mechanical models of intergenerational transmission. *The American Economic Review, 79*(3), 504–513.

18. Goodman, R. E., (1989). *Introduction to Rock Mechanics* (Vol. 2, pp. 1–15). New York: Wiley.

19. Graham, F. L., & Prevec, L., (1991). Manipulation of adenovirus vectors. In: *Gene Transfer and Expression Protocols* (Vol. 1, pp. 109–128). Humana press.

20. Grattan-Guinness, I., (1988). Mathematics, mechanics, and astronomy: Newton (1687), Lagrange (1788), Poincaré (1889). *Historia Mathematica, 15*(2), 165–170.

21. Guicciardini, N., (2005). Isaac Newton, philosophiae naturalis principia mathematica, (1687). In: *Landmark Writings in Western Mathematics 1640–1940* (Vol. 1, pp. 59–87). Elsevier Science.

22. Haber, R. B., (1990). Visualization techniques for engineering mechanics. *Computing Systems in Engineering, 1*(1), 37–50.

23. Habib, S., (1990). Classical limit in quantum cosmology: Quantum mechanics and the Wigner function. *Physical Review D, 42*(8), 2566.

24. Hepp, K., (1974). The classical limit for quantum mechanical correlation functions. *Communications in Mathematical Physics, 35*(4), 265–277.

25. Hibbeler, R. C., & Fan, S. C., (2004). *Statics and Mechanics of Materials* (Vol. 2, pp. 1–20). Singapore: Prentice-Hall.

26. Hinch, E. J., (1977). Mechanical models of dilute polymer solutions in strong flows. *The Physics of Fluids, 20*(10), S22–S30.

27. Inman, D. J., & Singh, R. C., (1994). *Engineering Vibration* (Vol. 3, pp. 1–14). Englewood Cliffs, NJ: Prentice-Hall.

28. Jian-Le, C., (2009). Conformal invariance and conserved quantities of Mei symmetry for general holonomic systems [J]. *Acta Physica. Sinica.*, *1(1), 1–10.*

29. Johansen, S., (1988). Statistical analysis of cointegration vectors. *Journal of Economic Dynamics and Control*, *12*(2/3), 231–254.

30. Jones, V., (1989). On knot invariants related to some statistical mechanical models. *Pacific Journal of Mathematics*, *137*(2), 311–334.

31. Karimi, M., Inzé, D., & Depicker, A., (2002). GATEWAY™ vectors for agrobacterium-mediated plant transformation. *Trends in Plant Science*, *7*(5), 193–195.

32. Kiros, R., Zhu, Y., Salakhutdinov, R. R., Zemel, R., Urtasun, R., Torralba, A., & Fidler, S., (2015). Skip-thought vectors. In: *Advances in Neural Information Processing Systems* (Vol. 1, pp. 3294–3302).

33. Kutz, M., (2015). *Mechanical Engineers' Handbook, Volume 1: Materials and Engineering Mechanics* (Vol. 1, pp. 1–30). John Wiley & Sons.

34. Lagomarsino, S., & Giovinazzi, S., (2006). Macroseismic and mechanical models for the vulnerability and damage assessment of current buildings. *Bulletin of Earthquake Engineering*, *4*(4), 415–443.

35. Leaf, B., (1968). Weyl transformation and the classical limit of quantum mechanics. *Journal of Mathematical Physics*, *9*(1), 65–72.

36. Lim, C. T., Zhou, E. H., & Quek, S. T., (2006). Mechanical models for living cells—a review. *Journal of Biomechanics*, *39*(2), 195–216.

37. Lister, J. R., & Kerr, R. C., (1991). Fluid-mechanical models of crack propagation and their application to magma transport in dykes. *Journal of Geophysical Research: Solid Earth*, *96*(B6), 10049–10077.

38. Lomov, S. V., Gusakov, A. V., Huysmans, G., Prodromou, A., & Verpoest, I., (2000). Textile geometry preprocessor for meso-mechanical models of woven composites. *Composites Science and Technology*, *60*(11), 2083–2095.

39. Ma, Q., & Zhang, Y., (2016). Mechanics of fractal-inspired horseshoe microstructures for applications in stretchable electronics. *Journal of Applied Mechanics*, *83*(11), 111008.

40. Man'ko, V. I., (1996). Classical formulation of quantum mechanics. *Journal of Russian Laser Research*, *17*(6), 579–584.

41. Maslov, V. P., & Fedoriuk, M. V., (2001). *Semi-Classical Approximation in Quantum Mechanics* (Vol. 7, pp. 1–15). Springer Science & Business Media.

42. McCormick, M. E., (2009). *Ocean Engineering Mechanics: With Applications* (Vol. 2, pp. 1–30). Cambridge University Press.

43. McIntyre, J., Zago, M., Berthoz, A., & Lacquaniti, F., (2001). Does the brain model Newton's laws? *Nature Neuroscience*, *4*(7), 693.

44. Meriam, J. L., & Kraige, L. G., (2012). *Engineering Mechanics: Dynamics* (Vol. 2, pp. 1–30). John Wiley & Sons.

45. Miller, A. D., (1992). Retroviral vectors. In: *Viral Expression Vectors* (Vol. 1, pp. 1–24). Springer, Berlin, Heidelberg.

46. Muhanna, R. L., & Mullen, R. L., (2001). Uncertainty in mechanics problems—Interval–based approach. *Journal of Engineering Mechanics*, *127*(6), 557–566.

47. Muhanna, R. L., Zhang, H., & Mullen, R. L., (2007). Interval finite elements as a basis for generalized models of uncertainty in engineering mechanics. *Reliable Computing*, *13*(2), 173–194.

48. Nelson, E., (1959). Analytic vectors. *Annals of Mathematics*, 1(1), 572–615.

49. Newton, I., (2002). *The Cambridge Companion to Newton* (Vol. 1, pp. 1–18). Cambridge University Press.

50. Ohayon, R., & Soize, C., (1997). *Structural Acoustics and Vibration: Mechanical Models, Variational Formulations and Discretization* (Vol. 1, pp. 2–30). Elsevier.

51. Pascale, R. T., Millemann, M., & Gioja, L., (2000). *Surfing the Edge of Chaos: The Laws of Nature and the New Laws of Business* (Vol. 1, pp. 10–18). Crown business.

52. Pennington, J., Socher, R., & Manning, C., (2014). Glove: Global vectors for word representation. In: *Proceedings of the 2014 Conference on Empirical Methods in Natural Language Processing (EMNLP)* (Vol. 1, pp. 1532–1543).

53. Phillips, A. W., (1950). *Mechanical Models in Economic Dynamics* (Vol. 1, pp. 283–305). London School of Economics and Political Science.

54. Robert, D., (1997). Semi-classical approximation in quantum mechanics. A survey of old and recent mathematical results. *Helvetica Physica. Acta*, *71*(1), 44–116.

55. Roe, P. L., (1981). Approximate Riemann solvers, parameter vectors, and difference schemes. *Journal of Computational Physics, 43*(2), 357–372.

56. Scheck, F., (2010). *Mechanics: From Newton's Laws to Deterministic Chaos* (Vol. 1, pp. 1–16). Springer Science & Business Media.

57. Streeter Jr, D. D., & Hanna, W. T., (1973). Engineering mechanics for successive states in canine left ventricular myocardium: I. cavity and wall geometry. *Circulation Research, 33*(6), 639–655.

58. Strocchi, F., (1966). Complex coordinates and quantum mechanics. *Reviews of Modern Physics, 38*(1), 36.

59. Suppes, P., (1966). The probabilistic argument for a non-classical logic of quantum mechanics. *Philosophy of Science, 33*(1/2), 14–21.

60. Vable, M., (1985). An algorithm based on the boundary element method for problems in engineering mechanics. *International Journal for Numerical Methods in Engineering, 21*(9), 1625–1640.

61. Verlinde, E., (2011). On the origin of gravity and the laws of Newton. *Journal of High Energy Physics, 2011*(4), 29.

62. Watkins, E., (1997). The laws of motion from Newton to Kant. *Perspectives on Science, 5,* 311–348.

63. White, B. Y., (1984). Designing computer games to help physics students understand Newton's laws of motion. *Cognition and Instruction, 1*(1), 69–108.

64. Yeo, S., & Zadnik, M., (2000). Newton, we have a problem. *Australian Science Teachers Journal, 46*(1), 9.

Work, Energy, and Power

CONTENTS

5.1. INTRODUCTION

Word energy is being used a lot in everyday life. Although it is often used quite loosely, it does have a very precise physical meaning. The amount of the capacity of something to do work is called energy. It is not a material substance. Energy can be measured and stored in other forms (Hubley and Wells, 1983; Gogotsi and Simon, 2011).

Although we usually hear people talking about energy consumption, energy is never destroyed. Energy is just converted from one form to another, doing work in the procedure. Some forms of energy are more useful to others than us—for instance, low-level heat energy. It is useful to comment about the withdrawal or consumption of energy resources, for instance, oil, coal, or wind, then the consumption of energy itself (Boyle, 2004; Lai and Nelson, 2007; Weaver et al., 2012).

The measurable amount of energy of a speeding bullet is associated with what is known as kinetic energy. The bullet has this energy because due to the change of chemical potential, stored in the gunpowder, makes work in the process which impulses the bullet. Due to the work done by a microwave oven, a hot cup of coffee has a measurable amount of thermal energy, which in turn grabbed electrical energy through the electrical grid (Wu et al., 2010; Miao et al., 2011; Infield and Freris, 2020).

In practice, there is always some loss to other forms of energy like heat and sound whenever work is done to move energy from one form to another. For instance, a typical light bulb is just about 3% capable to convert electrical energy to visible light, whereas a human being is about 25% capable to convert the chemical energy of food into work (Denholm and Margolis, 2007; Hall and Bain, 2008).

5.2. MEASUREMENT OF ENERGY AND WORK

The standard unit that is used in physics to measure work done and energy is the joule, which has the symbol J. In mechanics when a force of 1 Newton is applied to an object, 1 joule is the energy transmitted and moves it a distance of 1 meter (Ensher et al., 1996; Levine, 2005; Cole et al., 2012).

Another unit of energy is the Calorie. The amount of energy is written in Calories on the packet of an item of food (Zigler et al., 1991; Ural et al., 2003; Thomson et al., 2017). A common 60-gram chocolate bar, for instance, holds around 280 Calories of energy. One Calorie is defined as the amount of energy required to raise 1 kg of water by 1° Celsius. One Calorie is equal to 4184 joules, so one chocolate bar has 1.17 of stored energy MJ or 1.17 million joules. That's a lot of joules (Crooks, 1998; Branson and Johannigman, 2004; Danielsson et al., 2007).

5.3. WORK WAS DONE TO BURN OFF CALORIES

Suppose we're feeling mortified about eating a chocolate bar; we just want to find how much exercise we need to do to balance those additional 280 Calories (Benedict, 1910; Van Kleef et al., 2008; Nestle and Nesheim, 2012).

5.3.1. Pushing a Box

Let's have an easy method of exercise: pushing a heavyweight box in a room, see Figure 5.1.

Figure 5.1. A person pushes a box to the right.

Source: https://www.khanacademy.org/science/physics/work-and-energy/work-and-energy-tutorial/a/what-is-work.

By using a bathroom scale among ourselves and the box, we'll come to know that we can push through a force of 500 N. In the meantime, we use a measuring tape and stopwatch to measure our speed. This will be 0.25 meters per second (Mogilner and Oster, 2003; Hill and Peters, 2004; Foxcroft, 2012).

So to burn off the candy bar how much work do we need to do to the box? The definition of work, WWW, is underneath:

$$W = F \cdot \Delta x \qquad (1)$$

To burn off the energy in the candy bar he works we need is E = 280cal·4184 J/cal = 1.17MJ.

So, Δx, the distance, we need to move the box through is:

$$W = F \cdot \Delta x$$
$$1.17 \text{ MJ} = (500 \text{ N}) \cdot \Delta x$$
$$\frac{1.17 \times 10^6 \text{ J}}{500 \text{ N}} = \Delta x$$
$$2,340 \text{ m} = \Delta x$$

However, remember that the human body is about 25% capable to transfer stored energy from food into work. The actual energy we will balance is four times greater than the work done to the box. So, we just need to push the box over a distance of 585 m, which is just over five football fields long (Yamada, and Saito, 2001; Nemrava and Čermák, 2008; Kumar, 2015). Provided the known speed of 0.25 m/s that will take us:

585 m/0.25 m/s = 2340s

Exercise 5.1:

Assume that the force that we put on to the box, see Figure 5.1, is originally reduced but increases to a constant value as we warm up. For example, in the graph below we see that as the box is displaced more—i.e., x gets larger—the force, FFF, rises for the first 30 m, see Figure 5.2. How could we find the work done during the period where the force is changing?

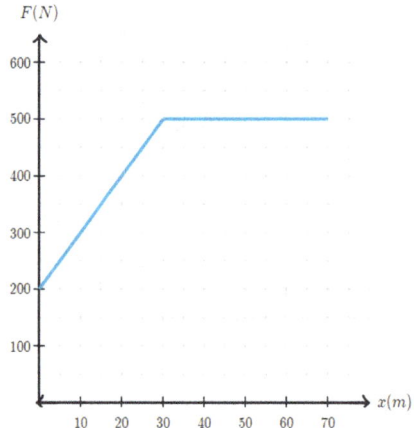

Figure 5.2. A varying force on the box.

Source: https://www.khanacademy.org/science/physics/work-and-energy/work-and-energy-tutorial/a/what-is-work.

When the force is not constant, work done can be determined by dividing the problem into the smaller sections through which the change is small and sum up the work done in each of the sections (Birkinshaw, 2010; Ahmad, 2013; Ahmad and Agah, 2014). We have just learned when looking at velocity-time graphs, by calculating the *area under the curve* by using geometry this can be done (Shibata and Iida, 2003; Parra-Gonzalez et al., 2009, 2011).

The work done via a force is equal to the area under a force vs. position graph. In the case of Figure 5.2, it would be:

(200 N·30 m) + 21 ((500 N−200 N)·30 m) = 10500J for the initial 30m of displacement. Similarly, the work done for the final 40 m of displacement would be:

500 N·40 m = 20,000 J

5.3.2. The Case of Non-Straight Pushing

When solving these problems there is one thing that we need to watch out for when doing these problems. The prior equation, $W = F \cdot \Delta x$, doesn't take into consideration in situations where the applying force is not in a similar direction as the motion (Van der Woude et al., 1995; Goosey et al., 1998; Mogilner and Oster, 2003). For example, imagine we are using a rope to pull on the box. In that situation, there will be an angle created between the ground and the rope. To disentangle this situation, we initiate by drawing a triangle to dispersed out the vertical and horizontal components of the applied force (Sanderson and Sommer III, 1985; De Looze et al., 2000; Hernández et al., 2014) (Figure 5.3).

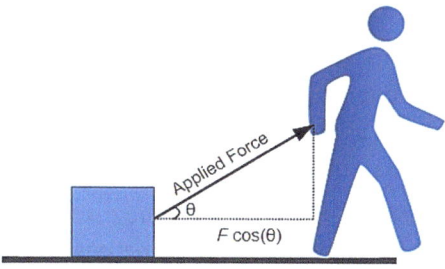

Figure 5.3. Force applied at a certain angle.

Source: https://www.khanacademy.org/science/physics/work-and-energy/work-and-energy-tutorial/a/what-is-work.

The main point here is that the force, $F\|$ is the only component which lies parallel to the work done on an object and the displacement. In the situation of the box shown beyond, work is being done by only the horizontal component of the applied force, $F\cos(\theta)$, on the box as the box is being moved horizontally (Hoozemans et al., 2007; Samozino et al., 2012; Russell et al., 2016). This means that a more common equation for the work done by a force on the box at an angle θ could be expressed as:

$W = F\| \cdot \Delta x$

$W = (F\cos\theta) \cdot \Delta x$

Which is more often written as,

$W = F\Delta x \cos\theta,$

5.3.3. Work Is Done by Lifting Weights

In the earlier example, we were pushing a box around a floor by doing work on it. In that case, we were working against a frictional force (Wong and Booth, 1988; Hadi et al., 2012).

Lifting weights is another common form of exercise. In this case, we are not working against the force of gravity but against the force of friction (Petrofsky and Lind, 978; Cogon et al., 1998). By using Newton's laws the force, F can be found which is essential to lift a weight with mass m straight up, just place it at a height h on a rack above us:

$F = mg$

The variation in position—earlier Δx—is simply the height, so the work, W, that we have done to lift the weight is then:

$W = mgh$

The result of the work done in lifting the weight is in the form of stored gravitational potential energy. Because it has the potential to be released at any moment with a crash as the weight falls back to the ground so it is called potential energy (Fenn, 1924; Gamberale, 1972; Garhammer, 1985).

As we have exerted our force in the same direction as the displacement of the weight, i.e., upwards, so we have done positive work on the weight. Since the force of gravity is directed towards the opposite direction to the displacement, so the work done by gravity was negative on the weight while it was lifted (Legwold and Kummant, 1982). Similarly, when the weight is stationary after the lift, the work that we have done is precisely canceled out due to the work done by gravity. The work done by gravity is $-mgh$, and the work done by us is *mgh*. We will discuss more this when we look into kinetic energy (Starr, 1951; Lin et al., 1999).

OK, let's set some numbers in and find how considerably of that chocolate bar we would balance by just lifting a weight of 50 kg to a height of 0.5 m. The work done on the weight is:

$$W = (50kg) (9.81m/s^2) (0.5m) = 245.25J$$

OK, so how many 1.17 x 106 joule—chocolate bars is this, i.e., 280 Calories? Fine, 245.25 J is about 1/4770 of one chocolate bar. But recall, our bodies are just 25% capable, so the work that a person does by is four times greater, around 981.8 J, which is 1/1190 chocolate bars. So, if for every 2 seconds we can lift this weight once, it will take us about 40 minutes or 2380 seconds of hard work to burn off this chocolate bar.

5.3.4. Holding a Weight Stationary

One common source of misunderstanding with the concept of work arises when thinking about lifting weight. Regardless any height lifted, we are not moving the weight it is returning to its initial position, so no work is being done (Garner et al., 1994; Duncan et al., 2010). We could also attain this by placing the load on a table; it is now clear that the table is not doing any sort of work to related to the weight position. However, we know from our experience that we will become tired doing the same job. So, what is going around here?

It turns out that, our bodies are doing work on our muscles to maintain the necessary tension to hold the weight. By sending a cascade of nerve impulses to each muscle, our body does this. Each impulse origins the muscle to temporarily release and contract (Lloyd and Zacks, 1972; Taylor, 1994; Lian et al., 1996). We just notice a slight trembling at first this indicates how fast it happens. Ultimately, however, not enough chemical energy is offered in the muscle and it can no longer keep up. We then instigate to jiggle and ultimately must rest for a while. So work is being done, it is nevertheless not being done on the weight (Harris et al., 1945; Wolkoff and Chilenskas, 1961; Osborne and Gilbert, 1980).

5.4. KINETIC ENERGY

The energy an object has due to its motion is called Kinetic energy. We must apply a force if we want to accelerate an object. to apply a force work must be done. Energy have to be transferred to the object after the work is done, and with a new constant speed (if the force is not acting anymore), the object will be in motion. This energy transferred is known as *kinetic energy*, and it depends on the speed achieved and mass (Viola et al., 1985; Agrawal et al., 1992; Skamarock, 2004).

Kinetic energy can be transformed into other kinds of energy and transferred among objects. For instance, a stationary chipmunk might collide with a flying squirrel. Succeeding the collision, some of the initial kinetic energy of the squirrel might have been transformed into some other form of energy or transferred into the chipmunk (Fjørtoft, 1953; Terray et al., 1996).

5.4.1. Calculation of Kinetic Energy

To calculate kinetic energy, we follow the reasoning above and initiate just by finding the work done, W, by a force, F, in an easy example. Let's assume a box of mass m being pushed over a distance d alongside a surface through a force parallel to that surface (Viola Jr, 1965). As we learned earlier:

W = F·d = m·a·d

If we remind our kinematic equations of motion, we'll get that we can substitute the acceleration if we know the final and initial velocity, vf, vi as well as the distance.

$$W = m \cdot d \cdot \frac{v_f^2 - v_i^2}{2d}$$
$$= m \cdot \frac{v_f^2 - v_i^2}{2}$$
$$= \frac{1}{2} \cdot m \cdot v_f^2 - \frac{1}{2} \cdot m \cdot v_i^2$$

So, once a remaining amount of work is done on an object, the quantity $1/2 \cdot m \cdot v^2$ which we call kinetic energy K—alternates.

Kinetic Energy: $K = 1/2 \cdot m \cdot v^2$

On the other hand, we can say that the variation (change) in kinetic energy is equal to the net work done on a system or object.

$W_{net} = \Delta K$

This outcome is famous as the work-energy theorem and applies quite commonly, even with forces that differ in magnitude and direction. It is vital in the study of conservation forces of and conservative energy (Muller-Dethlefs and Schlag, 1991; Mac et al., 1998).

5.4.2. Interesting Facts About Kinetic Energy

There are a few interesting things about kinetic energy that we can see from the equation:

- Kinetic energy relies on the velocity of the object squared. This means its kinetic energy quadruples when the velocity of an object doubles. A car moving at 60 mph has 4 times the kinetic energy of the same car traveling at 30 mph, and therefore the potential for 4 times higher destruction and death in the event of a crash (Sayvetz, 1939; Machlup and Onsager, 1953).

- Kinetic energy must all the time be either a positive value or zero. However, velocity can have a negative or positive value, the square of velocity is always positive.

- Kinetic energy is not a vector. Hence a tennis ball thrown *towards the right* with a velocity of 5 m/s, has the precise same kinetic energy as a tennis ball *thrown the down* with a velocity of 5 m/.

5.5. GRAVITATIONAL POTENTIAL ENERGY

We all know impulsively that a heavy load raised on somebody's head represents a *potentially* dangerous condition. The weight may be well protected, so there is no *essentially* dangerous. Our concern is that securing the weight against the gravity force is not really safe (Jones et al., 1996; Hatzfeld et al., 1997). To use correct physics vocabulary, we are worried about the *gravitational potential energy* of the weight.

The potential energy associated with all conservative forces. The force of gravity is no exclusion. *Ug* symbol is usually given to the gravitational potential energy. It signifies the potential an object has to do work as an outcome of being placed at a specific position in a gravitational field (Wolff, 1969; Ghosh et al., 2006).

Consider an object of mass m being lifted from a height h in contradiction of the force of gravity as shown underneath. The object is lifted vertically by a rope and pulley, so the force because of gravity, *Fg* the box and the force because of lifting, are parallel (Jones et al., 1998; Ghosh et al., 2009). If g is the magnitude (extent) of the gravitational acceleration, we can get the work done by the force on the weight just by multiplying the magnitude of the force of gravity, *Fg* times the vertical distance, *h*, it has moved over. This assumes the gravitational acceleration is continuous over the height *h* (Figure 5.4).

$Ug = Fg·h = m·g·h$

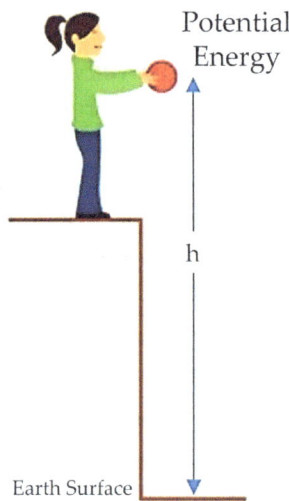

Figure 5.4. A weight lifted vertically to acquire gravitational potential energy.

Source: https://www.tes.com/lessons/PQ4Nky_diwdIrA/potential-and-kinetic-energy.

The object would fall back down to the ground if the force was to be removed and the gravitational potential energy would be changed into kinetic energy of the falling object. Our artifact on conservation of energy contains some example problems that are solved over knowledge of how gravitational potential energy is transformed into other forms (Manga and Kirchner, 2004; McKenzie and Jackson, 2012).

What is exciting about gravitational potential energy is that we can choose the zero arbitrarily. In other words, we have the freedom to select any vertical level as the location where $h = 0$. For simple mechanics problems, a suitable zero point would be at the surface of a table or on the floor of the laboratory. In principle nevertheless, we could select any orientation point— sometimes called a **datum**. If the object were to pass below the zero points, the gravitational potential energy could even be negative. This doesn't present a problem, however; we just have to be certain that the same zero points are used constantly in the calculation (Flesch et al., 2007; Schmalholz et al., 2014).

Exercise 5.2:

How much electrical energy would be used by an elevator lifting a 75 kg person through a height of 50 m if the elevator system has an overall efficiency of 25%? Assume the mass of the empty elevator car is properly balanced by a counterweight (Figure 5.5).

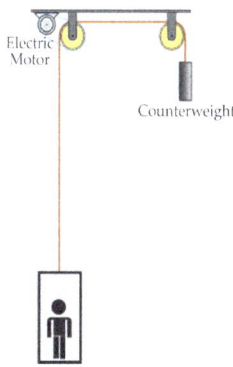

Figure 5.5. Schematic of an elevator system.

Source: https://www.khanacademy.org/science/physics/work-and-energy/work-and-energy-tutorial/a/what-is-gravitational-potential-energy.

Electrical energy through the grid is transferred into the gravitational potential energy of the heat and person because of friction inside the elevator system. There is no change in the gravitational potential energy of the elevator-counterweight system because the elevator car is counterweighted.

By using the equation for gravitational potential energy, we initially get the change in the gravitational potential energy of the person.

$$\Delta U_{\text{person}} = mgh$$
$$= (75 \text{ kg}) \cdot (9.81 \text{ m/s}^2) \cdot (50 \text{ m})$$
$$\simeq 3.68 \cdot 10^4 \text{ J}$$
$$\simeq 36.8 \text{ kJ}$$

We are told that there is an overall efficiency of 25% of the system. It means that 25% of the electric energy that is used by the motor is changed to useful work—gravitational potential energy in this case—and the residual 75% has vanished to the environment. So the total electrical energy use E is:

$$E = \frac{1}{0.25} \Delta U_{\text{person}}$$
$$= \frac{1}{0.25} 36.8 \text{ kJ}$$
$$= 147 \text{ kJ}$$

5.6. POTENTIAL ENERGY IN NON-UNIFORM GRAVITATIONAL FIELD

We'll no longer assume that the gravitational field is constant if the problem contains large distances. The attractive force between two masses, $m1$ and $m2$, decreases with separation distance r *squared* according to Newton's law of gravitation. If we take G as the gravitational constant,

$F = Gm_1 m_2 / r^2$

Typically, we choose the location of our zero points when dealing with gravitational potential energy over large distances which may seem counterintuitive. At a distance r of *infinity*, we place the zero point of gravitational potential energy (Isenberg and Kakad, 2009; Klotz, 2015). Due to which *all* values of the gravitational potential energy become negative.

As the distance r becomes large, the gravitational force tends rapidly towards zero so it makes sense to do this. When you are nearby to a planet, you are effectually bound to the planet through gravity and require a lot of energy to escape. You can only escape when $r = \infty$ but due to the inverse-square relationship, we can touch an asymptote where the gravitational potential energy turns out to be very close to zero. For a spacecraft departure earth, this can be supposed to happen at a height of about $10^{\wedge 7} \sim 5 \cdot 10^7$ meters above the external which is about 4 times the Earth's diameter. At that height, the acceleration has decreased to about 1% of the surface value because of gravity (Beck, 1960; Markx et al., 1997).

If we remember that work done is a force-time a distance then we can find that multiplying the force of gravity, above, with a distance r cancels out the squared in the denominator (Holliday and McIntyre, 1981; Webb et al., 2005). The gravitational potential energy is a function of r if we make our zero of potential energy at infinity:

$Ug\,(r) = -Gm_1m_2/\mathrm{r}$

This approach is very suitable for describing the energy requirements for traveling among different bodies in the solar system. We can assume landing on a planet. As we come nearer to the planet, we gain kinetic energy. Since energy is conserved, there would be decreased in gravitational potential energy to the justification for this—in other words, Ug becomes *more negative* (Longuet-Higgins and Stewart, 1961).

This picture guides to the idea of a *gravity well* to transfer from one planetary body to another you need to "climb out of" it. The image below displays a representation of the gravity wells of Pluto and its moon Charon, adjusted for a 1,000 kg spacecraft (Evans, 1979; Gubbins, 1980; Severne and Luwel, 1986) (Figure 5.6).

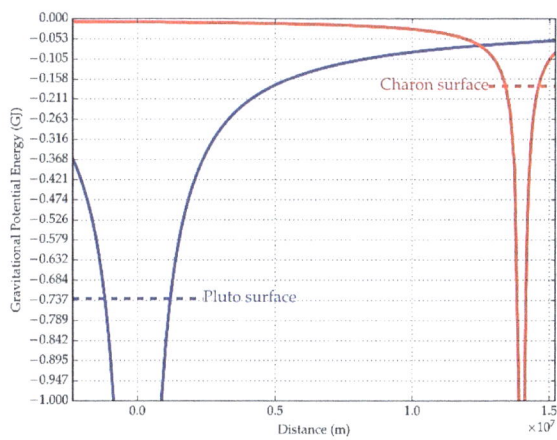

Figure 5.6. Gravitational potential energy wells of Pluto and Charon.

Source: https://www.khanacademy.org/science/physics/work-and-energy/ work-and-energy-tutorial/a/what-is-gravitational-potential-energy.

5.7. PRINCIPLE OF ENERGY CONSERVATION

The word conservation refers to something which doesn't change in physics. This means that the variable is constant above time in an equation which represents a conserved quantity. It has a similar value individually before and after an event (Driver and Warrington, 1985; Czarkowski and Kazimierczuk, 1993a, b).

In physics, there are many conserved quantities. They are often extraordinarily valuable for making predictions in what would otherwise be very complex conditions. In mechanics, three essential quantities are conserved. These are momentum, energy, and angular momentum (Haugan, 1979).

If you have observed the examples in other articles—for instance, the charging elephant's kinetic energy — energy is a conserved quantity. Nevertheless, in collisions, energy frequently changes (Tatar and Oktay, 2007). It turns out that there are a couple of important succeeding statements we need to add:

- Energy, in this measure, do not raises the total energy of a system. The energy associated with objects as they move around over time—e.g., potential, gravitational, kinetic, heat—might change forms, but if energy is conserved, then the total will stay the unchanged.

- Conservation of energy relates solitary to *isolated systems*. A ball rolling through a rough floor will not follow the law of conservation of energy since it is not separated (isolated) from the floor. The floor-through friction is in fact, doing work on the ball. Though, if we consider the floor and ball together, then conservation of energy will apply. We would generally call this combination the *ball-floor system*.

In mechanics problems, we are probable to meet systems having gravitational potential energy (*Ug*), kinetic energy (*EK*), elastic—spring—potential energy (*Us*), and thermal energy (heat) (*EH*). Solving such problems frequently initiates by establishing conservation of energy in a system among some initial time—subscript i—and at sometime later—subscript f.

$$E_{\text{Ki}} + U_{\text{gi}} + U_{\text{si}} = E_{\text{Kf}} + U_{\text{gf}} + U_{\text{sf}} + E_{\text{Hf}}$$

Which could be expanded out as:

$$\tfrac{1}{2}mv_i^2 + mgh_i + \tfrac{1}{2}kx_i^2 = \tfrac{1}{2}mv_f^2 + mgh_f + \tfrac{1}{2}kx_f^2 + E_{\text{Hf}}$$

5.7.1. Description of an Energy System

In physics, the *system* is the suffix we give to a gathering of objects that we pick to ideal with our equations. The system should contain the object of interest and *all* other objects that it cooperates with if we are to define the motion of an object by using the conservation of energy.

In practice, we always need to select to ignore some interactions. We are drawing a line around the things we care about and things we do not care about when defining a system. The things we do not comprise are generally cooperatively termed as *the environment*. Our calculations would be less accurate by ignoring some of the environment. There is no mortification in doing this though. Being a good physicist is often as much about knowing the effects you need to define as it is about understanding which effects can be carefully ignored (Bossel, 2003).

Consider a person who has the problem of making a bungee jump from a bridge. At least, the system should contain the bungee, jumper, and the Earth. A more precise calculation might contain the air, which does work on the jumper through air resistance, or drag. We could go more and include the bridge and its foundation, but because we know that the bridge is quite heavier than the jumper, we can carefully ignore this. We wouldn't imagine the force of a decelerating bungee jumper to have any major effect on the bridge, particularly if the bridge is intended to bear the load of heavy vehicles (Solbes et al., 2009; Chen et al., 2010) (Figure 5.7).

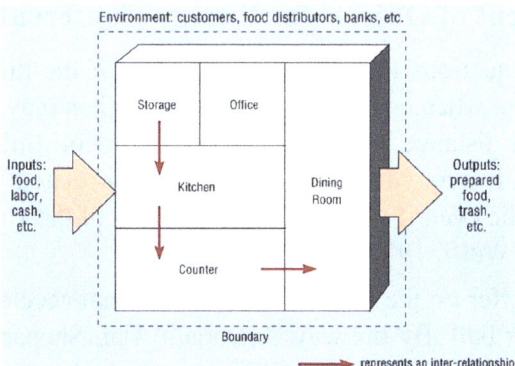

Figure 5.7. There is always some tenuous level of interaction between even distant objects, so we need to choose the boundary of our system intelligently (the dotted line represents the boundary).

There is always some weak level of interaction among even withdrawn objects, so we need to select the boundary of our system logically.

5.7.2. Mechanical Energy

Mechanical energy, E_M, is the sum of the kinetic energy and potential energy in a system.

$E_M = E_P + _{EK}$

Only gravity and the spring force that are conservative forces have potential energy-related with them. Nonconservative forces like drag and friction do not. Through a conservative force, we can always get back the energy that we put into a system. Energy transmitted by nonconservative forces though is tough to recover. It frequently ends up as heat or some other form that is normally lost to the environment—in other words, outside the system.

This means in practice is that the exceptional case of *conservation of mechanical energy* is frequently more advantageous for computing calculations than conservation of energy in common. Then all forces are conservative Conservation of mechanical energy only applies then. Fortunately, there are many circumstances where nonconservative forces are insignificant, or as a minimum, a good estimate can still be made when ignoring them.

5.7.3. Movement of Objects for Energy Conservation

We can set up equations that associate the sum of the different forms of energy in a system when energy is conserved. We then may be able to solve the equations for distance, velocity, or some other constraint on which the energy depends. It may still be beneficial to plot-related variables to see where solutions lie even if we don't know enough of the variables to find a unique solution (Wulff, 1974).

Imagine a golfer on the moon—the gravitational acceleration 1.625 m/s²—hitting a golf ball. By the way, Astronaut Alan Shepard truly did this. The ball exits the club at an angle of 45° towards the lunar surface traveling at 20 m/s both vertically and horizontally —total velocity 28.28 m/s. *How high would the golf ball go?*

We start by writing down the mechanical energy:

$E_M = 1/2 \ mv^2 + mgh$

By applying the principle of conservation of mechanical energy, we can resolve for the height h—note that the mass cancels out.

$$\tfrac{1}{2}mv_i^2 = mgh_f + \tfrac{1}{2}mv_f^2$$

$$
\begin{aligned}
h &= \frac{\tfrac{1}{2}v_i^2 - \tfrac{1}{2}v_f^2}{g} \\
&= \frac{\tfrac{1}{2}(28.28\ \mathrm{m/s})^2 - \tfrac{1}{2}(20\ \mathrm{m/s})^2}{1.625\ \mathrm{m/s^2}} \\
&= 123\ \mathrm{m}
\end{aligned}
$$

As you can see, applying the principle of conservation of energy permits us to rapidly solve problems comparable to this which would be tougher if done only with the kinematic equations.

5.7.4. Perpetual Motion Machines

A machine that keeps moving, working, changing internal energy to moves forever is known as the perpetual motion machine. An infinite variation of wonderful and weird machines has been defined over the years. They contain pumps used to run themselves through their head of falling water, wheels which are used to drive themselves around through unbalanced masses, and many differences of self-repelling magnets.

Although often interesting curious devices, a machine have never been shown to be perpetual, nor could it ever be. In fact, it wouldn't be very useful, even if such a machine were to exist, it would have no capability to do work. Note that this varies from the idea of the over-unity machine, which is used to output greater than 100% of the energy put into it, in clear violation of the principle of conservation of energy.

There is nothing that firmly makes the perpetual motion machine difficult according to the most basic principles of mechanics. The energy would be conserved and it would run forever if a system could be fully separated from the environment and subject to only conservative forces. The problem is that in reality, there is no technique to entirely separate a system and energy is never entirely conserved inside the machine.

In nowadays it is possible to make particularly low friction flywheels that rotate in a vacuum to store energy. However, they spin down when unloaded and still lose energy over some years (Quelhas et al., 2007). The rotation of the earth on its axis in space is perhaps an extreme example of such a machine. However, due to interactions with the moon, tidal friction, and other heavenly bodies, it too is increasingly slowing. Every couple of years, scientists need to add a *leap second* to our record of time to interpretation for the difference in the length of the day.

5.8. THERMAL ENERGY

The energy contained within a system that is responsible for its temperature known as *Thermal energy*. The Flow of thermal energy is called heat. A whole branch of *thermodynamics*, physics, deals with how heat is transferred among dissimilar systems and how work is done in the process (Hughes, 2012). From the perspective of mechanics problems, we are generally involved in the role thermal energy plays in guaranteeing the conservation of energy. Nearly every transfer of energy that takes place in real-world physical systems does so with competence lower than 100% and results in some thermal energy. This energy is typically in the form of *low-level* thermal energy. Now, low-level indicates that the temperature-related with thermal energy is near to that of the environment (Fishbone and Abilock, 1981). It is only probable to excerpt work when there is a temperature difference, thus low-level thermal energy signifies 'the end of the road' of energy transfer. No additional useful work is possible; the energy is currently 'lost to the environment.'

5.8.1. Thermal Energy from Friction

Imagine a man pushing a box through a rough floor at a constant velocity as shown in Figure 5.8. Because the friction force is non-conservative, the work done as potential energy is not stored. The work is done by the friction force outcomes in the transmission of energy into the thermal energy of the box-floor system. This thermal energy moves as heat inside the floor and box, finally raising the temperature of both of these objects (Hedström et al., 2004; Xia and Zhang, 2013).

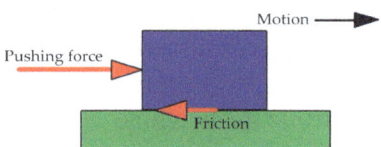

Figure 5.8. Man pushing a box opposed by friction.

Source: https://www.school-for-champions.com/science/friction_types.htm#. XcfVYFczayI.

By finding, the total work done by friction as the person pushes the box the change in total thermal energy ΔE of the box-floor system can be found. Remember that the box is moving at a constant velocity; this means that the applied force and the force of friction are equivalent in magnitude. The work done by both these forces is thus also equal (Manjarrez and Galvan, 1979; Wielicki et al., 1996).

By using the description of work done by a force parallel to the motion of an object moving over a distance d:

$W = F \cdot d$

$\Delta E_T = F_{friction} \cdot d$

If the coefficient of kinetic friction is kμk then this can also be written as:

$\Delta ET = \mu_k F_n d$

5.8.2. Thermal Energy from Dragging

The dragging force on a moving object because of a fluid like water or air is another example of a non-conservative force.

Some momentum is transferred, and the fluid is set in motion when an object moves over a fluid. There would still be some remaining motion of the fluid if the object were to stop moving. This would die down later sometime. The large-scale motions of the fluid are ultimately re-distributed into many minor random motions of the molecules in the fluid. These motions signify increased thermal energy in the system.

Figure 5.9 displays a system in which a thermally isolated water tank has a shaft deferred in it. There are two paddles are attached to the shaft which is established to rotate on its axis. In this system, any work is done in revolving the shaft effects in a transfer of kinetic energy to the water. There will still be some residual motion if the driving force is removed from the shaft after some time. Though, the motion will ultimately die down and increase the thermal energy of the water.

Remarkably, a system similar to that shown in Figure 5.9 was used by James Prescott Joule (1818–1889), for whom the SI unit of energy is termed. By using a paddle wheel immersed in a tank of whale oil and driven by falling weights, he was able to regulate the relationship between heat and mechanical energy. This guide to the 1st law of thermodynamics and law of conservation of energy.

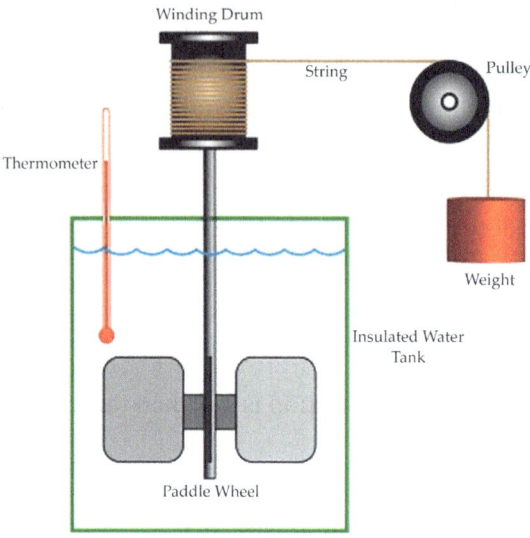

Figure 5.9. A paddlewheel rotating in a water tank.

Source: http://www.engineeringexpert.net/Engineering-Expert-Witness-Blog/ james-prescott-joule-and-the-joule-apparatus.

5.9. POWER

Alike *energy*, the *power* is a word we hear a lot. In daily life, it has an extensive range of meanings. In physics though, it has a very precise meaning. The measure of the rate at which work is done (or similarly, at which energy is transferred) is called power (Abhat, 1983; Hawlader et al., 2003).

The capability to precisely measure power was one of the important abilities which gave early engineers to make the steam engines which drove the industrial revolution. It remains to be important for knowing how to make usage of the energy resources which drive the modern world (Inaba, 2000; Sharma et al., 2009).

5.9.1. Measurement of Power

Watt is the standard unit used to measure power which has the symbol W. It was named after the Scottish inventor and industrialist James Watt. You have perhaps encountered the watt often in everyday life. The power output of electrical equipment like stereos or light bulbs is usually presented in watts (Zalba et al., 2003).

One watt is equal to one joule of work done per second by the definition. So if P_p denotes power in watts, Δt, is the time taken in seconds then and ΔE, is the change in energy (number of joules):

$$P = \frac{\Delta t}{\Delta E}$$

There is another unit of power that is quite extensively used: horsepower. This is generally given the symbol hp and has its roots in the 17th century where it referred to the power of a distinctive horse when being used to turn a capstan. Meanwhile then, a *metric horsepower* has been defined as the power compulsory to lift a 75 kg mass across a distance of 1 meter in 1 second. Thus, how much power is this in watts?

Well, we know that a mass attains gravitational potential energy, $Ep = m \cdot g \cdot h$ when being lifted against gravity. So putting in the numbers, we have:

$$\frac{75 \text{ kg} \cdot 9.807 \text{ m/s}^2 \cdot 1 \text{ m}}{1 \text{ s}} = 735.5 \text{ W}$$

5.9.2. Measurement Varying Power

The rate of usage differs over time in many conditions where energy resources are being used. The distinctive usage of electricity in a house is one such example. We see negligible usage throughout the day, afterward peaks when meals are ready, and a lengthy period of higher usage for evening heating and lighting (Figure 5.10) (Tian and Zhao, 2013).

There are at minimum three ways in which power is represented or expressed which are appropriate here: *peak power* P_{pk}, *Instantaneous power* Pi and *average power* P_{avg}. The electricity company needs to keep the way of all of these. Diverse energy resources are frequently transported to bear in addressing each of them.

- The power measured at a given instant in time is instantaneous power is. We'll get Δt extremely small if we consider the equation for power, $P = \Delta E / \Delta t$. If you are lucky enough to have a plot of time vs. power, the instantaneous power is the value you would read from the plot at any given time.

- The power measured over a long period is average power, i.e., when Δt in the equation for power is extremely large. One method to calculate this is to just find the area below the power vs. time curve (which gives the total work done) and divide by the total time. This is commonly best done with calculus, but it is frequently possible to estimate it sensibly precisely just by using geometry.

- The maximum value the instantaneous power can have in a specific system over a long period is known as the Peak power. Stereo systems and Car engines are an example of systems which can give a peak power which is much greater than their rated average power. Though, it is typically only possible to preserve this power for a short time if the damage is to be eluded. Yet, in these uses, a high peak power might be more vital to the listening or driving experience than high average power.

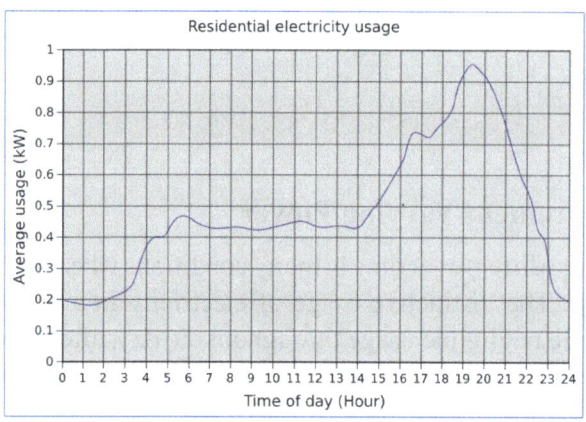

Figure 5.10

Source: https://www.khanacademy.org/science/physics/work-and-energy/work-and-energy-tutorial/a/what-is-power.

5.9.3. Description of Objects' Motion in Respect of Power

The equation for power joins time and work done. Because we know that work is done by forces, and forces can drift objects, we might suppose that knowing the power can allow us to understand something about the motion of a body over time.

By substituting the work done by a force, $W = F \cdot \Delta x \cos\theta$, into the equation for power $P = W/\Delta t$ we find:

$$P = \frac{F \cdot \Delta x \cdot \cos\theta}{\Delta t}$$

If the force is towards the direction of motion (as it is in many problems) then $\cos(\theta) = 1$ and the equation can be re-written:

$P = F \cdot v$

Velocity is a change in distance over time. Or consistently,

$Pi = m \cdot a \cdot v$

Note that in this particular equation we have made sure to identify that the power is the instantaneous power, Pi. This is since we have both velocity and acceleration in the equation and so the velocity is changing over time. It only makes sense if we take the velocity at a given instant. Then, we needed to use the average velocity, i.e.:

This can be a mainly advantageous result. Assume a car has a mass of 1000 kg and has a publicized power output to the wheels of 75 kW (around100 hp). The publicist claims that it has continuous acceleration over the range of 0–25 m/s.

By using only this information, we can find the time the car should take under perfect situations to accelerate from zero to a speed of 25 m/s.

$$P_{avg} = m \cdot a \cdot \frac{1}{2} v_{final}$$

Because acceleration is $\Delta v/\Delta t$:

$$P_{avg} = m \cdot (v_{final}/t) \cdot \frac{1}{2} v_{final}$$

$$= \frac{m v_{final}^2}{2t}$$

Which can be rearranged:

$$t = \frac{v_{final}^2 \cdot m}{2 \cdot P_{avg}}$$

$$= \frac{(25 \text{ m/s})^2 \cdot 1000 \text{ kg}}{2 \cdot 75000 \text{ W}}$$

$$= 4.17 \text{ s}$$

REFERENCES

1. Abhat, A., (1983). Low temperature latent heat thermal energy storage: Heat storage materials. *Solar Energy, 30*(4), 313–332.

2. Agrawal, Y. C., Terray, E. A., Donelan, M. A., Hwang, P. A., Williams, III, A. J., Drennan, W. M., & Krtaigorodskii, S. A., (1992). Enhanced dissipation of kinetic energy beneath surface waves. *Nature, 359*(6392), 219.

3. Ahmad, N., & Agah, A., (2014). Plan and intent recognition in a multi-agent system for collective box pushing. *Journal of Intelligent Systems, 23*(1), 95–108.

4. Ahmad, N., (2013). *Intent Recognition in Multi-Agent Systems: Collective Box Pushing and Cow Herding* (Vol. 1, pp. 2–49). Doctoral dissertation, University of Kansas).

5. Beck, A. E., (1960). An expanding earth with loss of gravitational potential energy. *Nature, 185*(4714), 677–678.

6. Benedict, F. G., (1910). The influence of mental and muscular work on nutritive processes. *Proceedings of the American Philosophical Society, 49*(195), 145–163.

7. Birkinshaw, J., (2010). *Reinventing Management: Smarter Choices for Getting Work Done* (Vol. 1, pp. 2–19). John Wiley & Sons.

8. Bossel, U., (2003). Well-to-wheel studies, heating values, and the energy conservation principle. In: *European Fuel Cell Forum* (Vol. 22, pp. 1–5).

9. Boyle, G., (2004). Renewable energy. In: Godfrey, B., (ed.), *Renewable Energy* (Vol. 1, No. 1, p. 456). Oxford University Press, ISBN-10:0199261784; ISBN-13:9780199261789.

10. Branson, R. D., & Johannigman, J. A., (2004). The measurement of energy expenditure. *Nutrition in Clinical Practice, 19*(6), 622–636.

11. Chen, Q., Wang, M., & Guo, Z. Y., (2010). Field synergy principle for energy conservation analysis and application. *Advances in Mechanical Engineering, 2*, 129313.

12. Cogon, D., Kellingray, S., Inskip, H., Croft, P., Campbell, L., & Cooper, C., (1998). Osteoarthritis of the hip and occupational lifting. *American Journal of Epidemiology, 147*(6), 523–528.

13. Cole, M. S., Bruch, H., & Vogel, B., (2012). Energy at work: A measurement validation and linkage to unit effectiveness. *Journal of*

Organizational Behavior, *33*(4), 445–467.

14. Crooks, G. E., (1998). Nonequilibrium measurements of free energy differences for microscopically reversible Markovian systems. *Journal of Statistical Physics*, *90*(5/6), 1481–1487.

15. Czarkowski, D., & Kazimierczuk, M. K., (1993a). Energy-conservation approach to modeling PWM DC-DC converters. *IEEE Transactions on Aerospace and Electronic Systems*, *29*(3), 1059–1063.

16. Danielsson, A., Willén, C., & Sunnerhagen, K. S., (2007). Measurement of energy cost by the physiological cost index in walking after stroke. *Archives of Physical Medicine and Rehabilitation*, *88*(10), 1298–1303.

17. De Looze, M. P., Van Greuningen, K., Rebel, J., Kingma, I., & Kuijer, P. P. F. M., (2000). Force direction and physical load in dynamic pushing and pulling. *Ergonomics*, *43*(3), 377–390.

18. Denholm, P., & Margolis, R. M., (2007). Evaluating the limits of solar photovoltaics (PV) in electric power systems utilizing energy storage and other enabling technologies. *Energy Policy*, *35*(9), 4424–4433.

19. Driver, R., & Warrington, L., (1985). Students' use of the principle of energy conservation in problem situations. *Physics Education*, *20*(4), 171–76.

20. Duncan, C. A., MacKinnon, S. N., & Albert, W. J., (2010). Changes in thoracolumbar kinematics and centre of pressure when performing stationary tasks in moving environments. *International Journal of Industrial Ergonomics*, *40*(6), 648–654.

21. Ensher, J. R., Jin, D. S., Matthews, M. R., Wieman, C. E., & Cornell, E. A., (1996). Bose-Einstein condensation in a dilute gas: Measurement of energy and ground-state occupation. *Physical Review Letters*, *77*(25), 4984.

22. Evans, R., (1979). The nature of the liquid-vapor interface and other topics in the statistical mechanics of non-uniform, classical fluids. *Advances in Physics*, *28*(2), 143–200.

23. Fenn, W. O., (1924). The relation between the work performed and the energy liberated in muscular contraction. *The Journal of Physiology*, *58*(6), 373–395.

24. Fishbone, L. G., & Abilock, H., (1981). Markal, a linear-programming model for energy systems analysis: Technical description of the BNL version. *International Journal of Energy Research*, *5*(4), 353–375.

25. Fjørtoft, R., (1953). On the changes in the spectral distribution of

kinetic energy for two dimensional, nondivergent flow. *Tellus*, *5*(3), 225–230.

26. Flesch, L. M., Holt, W. E., Haines, A. J., Wen, L., & Shen-Tu, B., (2007). The dynamics of western North America: Stress magnitudes and the relative role of gravitational potential energy, plate interaction at the boundary and basal tractions. *Geophysical Journal International*, *169*(3), 866–896.

27. Foxcroft, L., (2012). *Calories and Corsets: A History of Dieting Over Two Thousand Years* (Vol. 1, pp. 2–16). Profile Books.

28. Gamberale, F., (1972). Perceived exertion, heart rate, oxygen uptake and blood lactate in different work operations. *Ergonomics*, *15*(5), 545–554.

29. Garhammer, J., (1985). Biomechanical profiles of Olympic weightlifters. *Journal of Applied Biomechanics*, *1*(2), 122–130.

30. Garner, C., Brophy, J., Polk, J., & Pless, L., (1994). Cyclic endurance test of a SPT-100 stationary plasma thruster. In: *30th Joint Propulsion Conference and Exhibit* (Vol. 1, pp. 2840–2856).

31. Ghosh, A., Holt, W. E., & Flesch, L. M., (2009). Contribution of gravitational potential energy differences to the global stress field. *Geophysical Journal International*, *179*(2), 787–812.

32. Ghosh, A., Holt, W. E., Flesch, L. M., & Haines, A. J., (2006). Gravitational potential energy of the Tibetan plateau and the forces driving the Indian plate. *Geology*, *34*(5), 321–324.

33. Gogotsi, Y., & Simon, P., (2011). True performance metrics in electrochemical energy storage. *Science*, *334*(6058), 917–918.

34. Goosey, V. L., Campbell, I. G., & Fowler, N. E., (1998). The relationship between three-dimensional wheelchair propulsion techniques and pushing economy. *Journal of Applied Biomechanics*, *14*(4), 412–427.

35. Gubbins, K. E., (1980). Structure of non-uniform molecular fluids: Integrodifferential equations for the density-orientation profile. *Chemical Physics Letters*, *76*(2), 329–332.

36. Hadi, G., Akkus, H., & Harbili, E., (2012). Three-dimensional kinematic analysis of the snatch technique for lifting different barbell weights. *The Journal of Strength and Conditioning Research*, *26*(6), 1568–1576.

37. Hall, P. J., & Bain, E. J., (2008). Energy-storage technologies and electricity generation. *Energy Policy*, *36*(12), 4352–4355.

38. Harris, E. E., Beglinger, E., Hajny, G. J., & Sherrard, E. C., (1945). Hydrolysis of wood-treatment with sulfuric acid in a stationary digester. *Industrial and Engineering Chemistry, 37*(1), 12–23.

39. Hatzfeld, D., Martinod, J., Bastet, G., & Gautier, P., (1997). An analog experiment for the Aegean to describe the contribution of gravitational potential energy. *Journal of Geophysical Research: Solid Earth, 102*(B1), 649–659.

40. Haugan, M. P., (1979). Energy conservation and the principle of equivalence. *Annals of Physics, 118*(1), 156–186.

41. Hawlader, M. N. A., Uddin, M. S., & Khin, M. M., (2003). Microencapsulated PCM thermal-energy storage system. *Applied Energy, 74*(1/2), 195–202.

42. Hedström, L., Wallmark, C., Alvfors, P., Rissanen, M., Stridh, B., & Ekman, J., (2004). Description and modeling of the solar-hydrogen-biogas-fuel cell system in Glashusette. *Journal of Power Sources, 131*(1/2), 340–350.

43. Hernández, G. E., Criswell, B. A., Kirk, N. J., Sauder, D. G., & Rushton, G. T., (2014). Pushing for particulate level models of adiabatic and isothermal processes in upper-level chemistry courses: A qualitative study. *Chemistry Education Research and Practice, 15*(3), 354–365.

44. Hill, J. O., & Peters, J. C., (2004). *The Step Diet Book: Count Steps, Not Calories to Lose Weight and Keep it Off Forever* (Vol. 1, pp. 2–26). Workman Publishing.

45. Holliday, D., & Mcintyre, M. E., (1981). On potential energy density in an incompressible, stratified fluid. *Journal of Fluid Mechanics, 107*, 221–225.

46. Hoozemans, M. J., Slaghuis, W., Faber, G. S., & van Dieen, J. H., (2007). Cart pushing: The effects of magnitude and direction of the exerted push force, and of trunk inclination on low back loading. *International Journal of Industrial Ergonomics, 37*(11/12), 832–844.

47. Hubley, C. L., & Wells, R. P., (1983). A work-energy approach to determine individual joint contributions to vertical jump performance. *European Journal of Applied Physiology and Occupational Physiology, 50*(2), 247–254.

48. Hughes, L., (2012). A generic framework for the description and analysis of energy security in an energy system. *Energy Policy, 42*, 221–231.

49. Inaba, H., (2000). New challenge in advanced thermal energy transportation using functionally thermal fluids. *International Journal of Thermal Sciences, 39*(9–11), 991–1003.

50. Infield, D., & Freris, L., (2020). *Renewable Energy in Power Systems* (Vol. 1, pp. 1–12). John Wiley & Sons.

51. Isenberg, D. R., & Kakad, Y. P., (2009). Simulating the effects of a non-uniform gravitational field on a space robot. *JCP, 4*(12), 1255–1262.

52. Jones, C. H., Sonder, L. J., & Unruh, J. R., (1998). Lithospheric gravitational potential energy and past orogenesis: Implications for conditions of initial basin and range and laramide deformation. *Geology, 26*(7), 639–642.

53. Jones, C. H., Unruh, J. R., & Sonder, L. J., (1996). The role of gravitational potential energy in active deformation in the southwestern United States. *Nature, 381*(6577), 37.

54. Kazimierczuk, M. K., & Czarkowski, D., (1993b). Application of the principle of energy conservation to modeling the PWM converters. In: *Proceedings of IEEE International Conference on Control and Applications* (Vol. 1, pp. 291–296). IEEE.

55. Klotz, A. R., (2015). The gravity tunnel in a non-uniform Earth. *American Journal of Physics, 83*(3), 231–237.

56. Kumar, G. S., (2015). Box pushing technique on railway under bridge for cross traffic works. *International Journal and Magazine of Engineering, Technology, Management and Research, ISSN,* (2320–3706), 5.

57. Lai, J. S., & Nelson, D. J., (2007). Energy management power converters in hybrid electric and fuel cell vehicles. *Proceedings of the IEEE, 95*(4), 766–777.

58. Legwold, G., & Kummant, I., (1982). Does lifting weights harm a prepubescent athlete? *The Physician and Sports Medicine, 10*(7), 141–144.

59. Levine, J. A., (2005). Measurement of energy expenditure. *Public Health Nutrition, 8*(7a), 1123–1132.

60. Lian, Ø., Engebretsen, L., Øvrebø, R. V., & Bahr, R., (1996). Characteristics of the leg extensors in male volleyball players with jumper's knee. *The American Journal of Sports Medicine, 24*(3), 380–385.

61. Lin, C. J., Bernard, T. M., & Ayoub, M. M., (1999). A biomechanical

evaluation of lifting speed using work-and moment-related measures. *Ergonomics*, *42*(8), 1051–1059.

62. Lloyd, B. B., & Zacks, R. M., (1972). The mechanical efficiency of treadmill running against a horizontal impeding force. *The Journal of Physiology*, *223*(2), 355–363.

63. Longuet-Higgins, M. S., & Stewart, R. W., (1961). The changes in amplitude of short gravity waves on steady non-uniform currents. *Journal of Fluid Mechanics*, *10*(4), 529–549.

64. Mac Low, M. M., Klessen, R. S., Burkert, A., & Smith, M. D., (1998). Kinetic energy decay rates of supersonic and super-Alfvénic turbulence in star-forming clouds. *Physical Review Letters*, *80*(13), 2754.

65. Machlup, S., & Onsager, L., (1953). Fluctuations and irreversible process. II. Systems with kinetic energy. *Physical Review*, *91*(6), 1512.

66. Manga, M., & Kirchner, J. W., (2004). Interpreting the temperature of water at cold springs and the importance of gravitational potential energy. *Water Resources Research*, *40*(5), 3–19.

67. Manjarrez, R., & Galvan, M., (1979). Solar multistage flash evaporation (SMSF) as a solar energy application on desalination processes. Description of one demonstration project. *Desalination*, *31*(1–3), 545–554.

68. Markx, G. H., Rousselet, J., & Pethig, R., (1997). DEP-FFF: Field-flow fractionation using non-uniform electric fields. *Journal of Liquid Chromatography and Related Technologies*, *20*(16/17), 2857–2872.

69. McKenzie, D., & Jackson, J., (2012). Tsunami earthquake generation by the release of gravitational potential energy. *Earth and Planetary Science Letters*, *345*, 1–8.

70. Miao, G., Himayat, N., Li, G. Y., & Talwar, S., (2011). Distributed interference-aware energy-efficient power optimization. *IEEE Transactions on Wireless Communications*, *10*(4), 1323–1333.

71. Mogilner, A., & Oster, G., (2003). Polymer motors: Pushing out the front and pulling up the back. *Current Biology*, *13*(18), R721–R733.

72. Muller-Dethlefs, K., & Schlag, E. W., (1991). High-resolution zero kinetic energy (ZEKE) photoelectron spectroscopy of molecular systems. *Annual Review of Physical Chemistry*, *42*(1), 109–136.

73. Nemrava, M., & Čermák, P., (2008). Solving the box-pushing problem by master-slave robots cooperation. *Journal of Automation Mobile Robotics and Intelligent Systems*, *2*, 32–37.

74. Nestle, M., & Nesheim, M., (2012). *Why Calories Count: From Science to Politics* (Vol. 33, pp. 1–12). University of California Press.

75. Osborne, R. J., & Gilbert, J. K., (1980). A method for investigating concept understanding in science. *European Journal of Science Education, 2*(3), 311–321.

76. Parra-Gonzalez, E. F., & Ramírez-Torres, J. G., (2011). Object path planner for the box pushing problem. *Multi-Robot Systems, Trends and Development, In Tech Pub,* 1(1), 291–306.

77. Parra-Gonzalez, E. F., Ramirez-Torres, J. G., & Toscano-Pulido, G., (2009). A new object path planner for the box pushing problem. In: *2009 Electronics, Robotics and Automotive Mechanics Conference (CERMA)* (Vol. 1, pp. 119–124). IEEE.

78. Petrofsky, J. S., & Lind, A. R., (1978). Comparison of metabolic and ventilatory responses of men to various lifting tasks and bicycle ergometry. *Journal of Applied Physiology, 45*(1), 60–63.

79. Quelhas, A., Gil, E., McCalley, J. D., & Ryan, S. M., (2007). A multiperiod generalized network flow model of the US integrated energy system: Part I—Model description. *IEEE Transactions on Power Systems, 22*(2), 829–836.

80. Russell Esposito, E., Aldridge Whitehead, J. M., & Wilken, J. M., (2016). Step-to-step transition work during level and inclined walking using passive and powered ankle–foot prostheses. *Prosthetics and Orthotics International, 40*(3), 311–319.

81. Samozino, P., Rejc, E., Di Prampero, P. E., Belli, A., & Morin, J. B., (2012). Optimal force–velocity profile in ballistic movements—Altius. *Medicine and Science in Sports and Exercise, 44*(2), 313–322.

82. Sanderson, D. J., & Sommer, III, H. J., (1985). Kinematic features of wheelchair propulsion. *Journal of Biomechanics, 18*(6), 423–429.

83. Sayvetz, A., (1939). The kinetic energy of polyatomic molecules. *The Journal of Chemical Physics, 7*(6), 383–389.

84. Schmalholz, S. M., Medvedev, S., Lechmann, S. M., & Podladchikov, Y., (2014). Relationship between tectonic overpressure, deviatoric stress, driving force, isostasy and gravitational potential energy. *Geophysical Journal International, 197*(2), 680–696.

85. Severne, G., & Luwel, M., (1986). Violent relaxation and mixing in non-uniform one-dimensional gravitational systems. *Astrophysics and Space Science, 122*(2), 299–325.

86. Sharma, A., Tyagi, V. V., Chen, C. R., & Buddhi, D., (2009). Review on thermal energy storage with phase change materials and applications. *Renewable and Sustainable Energy Reviews, 13*(2), 318–345.

87. Shibata, K., & Iida, M., (2003). Acquisition of box pushing by direct-vision-based reinforcement learning. In: *SICE 2003 Annual Conference (IEEE Cat. No. 03TH8734)* (Vol. 3, pp. 2322–2327). IEEE.

88. Skamarock, W. C., (2004). Evaluating mesoscale NWP models using kinetic energy spectra. *Monthly Weather Review, 132*(12), 3019–3032.

89. Solbes, J., Guisasola, J., & Tarín, F., (2009). Teaching energy conservation as a unifying principle in physics. *Journal of Science Education and Technology, 18*(3), 265–274.

90. Starr, I., (1951). Units for the expression of both static and dynamic work in similar terms, and their application to weight-lifting experiments. *Journal of Applied Physiology, 4*(1), 21–29.

91. Tatar, E., & Oktay, M., (2007). Students' misunderstandings about the energy conservation principle: A general view to studies in literature. *International Journal of Environmental and Science Education, 2*(3), 79–81.

92. Taylor, M. A., (1994). Stone, bone or blubber? Buoyancy control strategies in aquatic tetrapods. *Mechanics and Physiology of Animal Swimming, 1*(1), 151–161.

93. Terray, E. A., Donelan, M. A., Agrawal, Y. C., Drennan, W. M., Kahma, K. K., Williams, A. J., & Kitaigorodskii, S. A., (1996). Estimates of kinetic energy dissipation under breaking waves. *Journal of Physical Oceanography, 26*(5), 792–807.

94. Thomson, H., Bouzarovski, S., & Snell, C., (2017). Rethinking the measurement of energy poverty in Europe: A critical analysis of indicators and data. *Indoor and Built Environment, 26*(7), 879–901.

95. Tian, Y., & Zhao, C. Y., (2013). A review of solar collectors and thermal energy storage in solar thermal applications. *Applied Energy, 104*, 538–553.

96. Ural, A., Zehnder, A. T., & Ingraffea, A. R., (2003). Fracture mechanics approach to face sheet delamination in honeycomb: Measurement of energy release rate of the adhesive bond. *Engineering Fracture Mechanics, 70*(1), 93–103.

97. Van der Woude, L. H. V., Van Konlngsbruggen, C. M., Kroes, A. L., & Kingma, I., (1995). Effect of push handle height on net moments and

forces on the musculoskeletal system during standardized wheelchair pushing tasks. *Prosthetics and Orthotics International, 19*(3), 188–201.

98. Van Kleef, E., Van Trijp, H., Paeps, F., & Fernandez-Celemin, L., (2008). Consumer preferences for front-of-pack calories labeling. *Public Health Nutrition, 11*(2), 203–213.

99. Viola Jr, V. E., (1965). Correlation of fission fragment kinetic energy data. *Nuclear Data Sheets. Section A, 1*, 391–410.

100. Viola, V. E., Kwiatkowski, K., & Walker, M., (1985). Systematics of fission fragment total kinetic energy release. *Physical Review C, 31*(4), 1550.

101. Weaver, V. M., Johnson, M., Kasichayanula, K., Ralph, J., Luszczek, P., Terpstra, D., & Moore, S., (2012). Measuring energy and power with PAPI. In: *2012 41st International Conference on Parallel Processing Workshops* (Vol. 1, pp. 262–268). IEEE.

102. Webb, G. M., Zank, G. P., Kaghashvili, E. K., & Ratkiewicz, R. E., (2005). Magnetohydrodynamic waves in non-uniform flows II: Stress-energy tensors, conservation laws, and lie symmetries. *Journal of Plasma Physics, 71*(6), 811–857.

103. Wielicki, B. A., Barkstrom, B. R., Harrison, E. F., Lee, III, R. B., Smith, G. L., & Cooper, J. E., (1996). Clouds and the earth's radiant energy system (CERES): An earth observing system experiment. *Bulletin of the American Meteorological Society, 77*(5), 853–868.

104. Wolff, M., (1969). Direct measurements of the Earth's gravitational potential using a satellite pair. *Journal of Geophysical Research, 74*(22), 5295–5300.

105. Wolkoff, J., & Chilenskas, A. A., (1961). The melt refining of irradiated uranium: Application to EBR-II fast reactor fuel. IX. Sorption and retention of sodium and cesium vapor on stationary beds at elevated temperature. *Nuclear Science and Engineering, 9*(1), 71–77.

106. Wong, T. S., & Booth, F. W., (1988). Skeletal muscle enlargement with weight-lifting exercise by rats. *Journal of Applied Physiology, 65*(2), 950–954.

107. Wu, D., Aliprantis, D. C., & Gkritza, K., (2010). Electric energy and power consumption by light-duty plug-in electric vehicles. *IEEE Transactions on Power Systems, 26*(2), 738–746.

108. Wulff, W., (1974). The energy conservation equation for living tissue. *IEEE Transactions on Biomedical Engineering*, (6), 494–495.

109. Xia, X., & Zhang, J., (2013). Mathematical description for the measurement and verification of energy efficiency improvement. *Applied Energy, 111*, 247–256.

110. Yamada, S., & Saito, J. Y., (2001). Adaptive action selection without explicit communication for multirobot box-pushing. *IEEE Transactions on Systems, Man, and Cybernetics, Part C (Applications and Reviews), 31*(3), 398–404.

111. Zalba, B., Marın, J. M., Cabeza, L. F., & Mehling, H., (2003). Review on thermal energy storage with phase change: Materials, heat transfer analysis and applications. *Applied Thermal Engineering, 23*(3), 251–283.

112. Zigler, A., Burkhalter, P. G., Nagel, D. J., Rosen, M. D., Boyer, K., Gibson, G., & Rhodes, C. K., (1991). Measurement of energy penetration depth of subpicosecond laser energy into solid density matter. *Applied Physics Letters, 59*(5), 534–536.

Structure of Materials

CONTENTS

6.1. INTRODUCTION

It must be clear that all matter is composed of atoms. From the periodic table (Figure 6.1), it can be viewed that there are only roughly 100 different types of known atoms in the whole Universe (some new elements are known and identified in laboratories, up to Z=116). These same hundred atoms produce thousands of diverse substances varying from the air to metal used to aid tall buildings. Metals behave differently as compared to ceramics and ceramics behave in a different way than the polymers (Sutton, 1993; De Graef and McHenry, 2012). The properties of the matter are dependent on which kind of atoms are utilized and how these atoms are bonded together. The structure of the materials can be categorized by the magnitude of several features being considered. The three most usual major classifications of the structure are given below:

The atomic structure comprises features that can't be seen, like types of bonding amongst the atoms and the way those atoms are arranged. Microstructure comprises features that generally can be viewed using the microscope, but rarely with the naked eye. Macrostructure comprises features that usually can be viewed with the naked eye (Larson, 1999; Smith et al., 2006; Wong and Salleo, 2009).

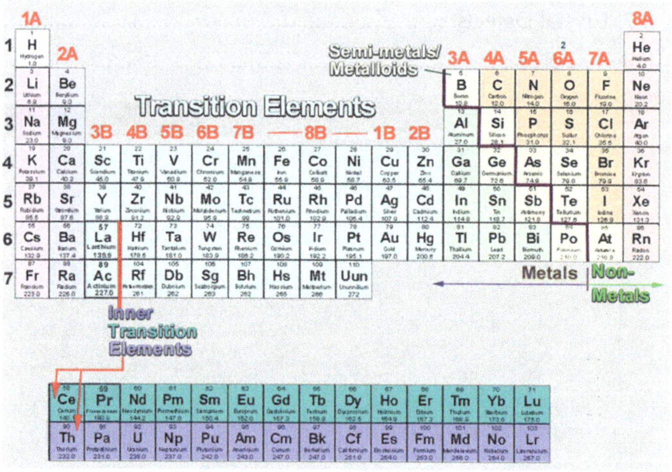

Figure 6.1. Periodic table of the elements.

Source: https://www.nde-ed.org/EducationResources/CommunityCollege/Materials/Structure/introduction.htm.

The atomic structure mainly affects the thermal, physical, chemical, electrical, optical, and magnetic properties (Newnham, 2005; Fratzl, 2008). The macrostructure and microstructure can affect these properties but generally, they have larger effect on the mechanical properties and on rate of the chemical reaction. The properties of the material provide hints as to the structure of material (McKeown, 1998; Meyers et al., 2008). The strength of the metals recommends that the atoms are bonded together by strong bonds. However, these bonds should also permit atoms to move as metals are also normally formable. To comprehend the structure of the material, the kind of atoms existent, and how these atoms are bonded and arranged must be known (Hüsing and Schubert, 1998; Ehrenstein, 2012).

6.2. ATOMIC BONDS

From the elementary chemistry, it is already known that atomic structure of the element is composed of positively charged nucleus encircled by the electrons revolving around it. The atomic number of an element specifies the number of positively charged protons in nucleus. An atom's atomic weight specifies how many neutrons and protons are in the nucleus. To find out the number of the neutrons in an atom, atomic number is subtracted simply from the atomic weight (Tanaka et al., 1999; Evers et al., 2015).

Atoms tends to possess a balanced electrical charge. Thus, they generally comprise negatively charged electrons encircling the nucleus in the numbers equal to number of the protons. It is also already known that the electrons are existent with dissimilar energies and it is appropriate to contemplate these electrons encircling the nucleus in the energy shells (Welker and Giessibl, 2012). For instance, magnesium, having an atomic number of twelve, has two electrons in its inner shell, 8 in second shells and two in the outer shell (Megaw, 1958; Volokh, 2008).

All chemical bonds include electrons. Atoms will stay nearby to each other if they have a mutual interest in the one or more than one electrons. Atoms are most stable when they possess no partially-filled electron shells. If an atom possesses only a few electrons in its shell, it will incline to lose these electrons to vacant the shell (Wrinch, and Harker, 1940; Dmitriev et al., 2004). These elements are metals. When the metal atoms bond, the metallic bond occurs. When an atom has almost full electron shell, the atom will try to discover electrons from the other atom in order to fill its outer shell. These elements are normally described as nonmetals. The bond amongst the two non-metal atoms is generally the covalent bond. Where non-metal and metal atom comes close together, an ionic bond takes place. There are also some other, less common, kinds of bond but the information is beyond the room of this material (Ono and Koyanagi, 2000; Lawniczak-Jablonska et al., 2002) (Figure 6.2).

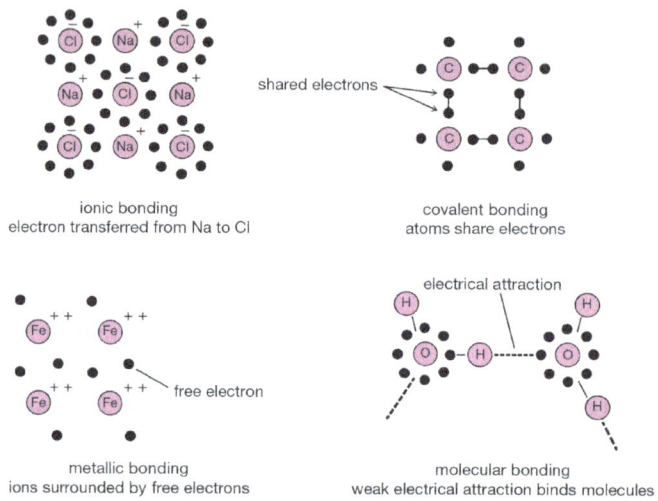

Figure 6.2. Different kinds of atomic bonds.

Source: https://www.britannica.com/science/chemical-bonding.

6.3. SOLID-STATE STRUCTURE

In the previous section, some mechanisms which bond together the gathering of individual molecules or atoms of the solid material were described (Klok and Lecommandoux, 2006; Xia et al., 2012). These forces might be key chemical bonds, as in the metals and ionic solids, or van der Waals' forces of the solids, just like in paraffin wax, ice, and most of the polymers. In solids, the manner the molecules or atoms arrange themselves donates to the presence and properties of the materials (Johnston et al., 1995; Prashanth et al., 2002).

Atoms can be grouped together as a collection through the number of different procedures, including condensation, chemical reaction, pressurization, melting, and electrodeposition. The process normally determines, initially as a minimum, whether the gathering of atoms will occur to form a solid, liquid, or gas. The state normally changes as its pressure or temperature is changed. Melting is the procedure most frequently used to create an aggregate of the atoms (Haubold et al., 1989; Mas-Torrent and Rovira, 2011). When the temperature of the melt is lowered to a particular point, the liquid will either form a crystalline solid or the amorphous solid (Jutzi et al., 1991; Sheth et al., 2003).

6.3.1. Amorphous Solids

The solid substance having the atoms held apart at the equilibrium configuration but having no long-range periodicity in the location of an atom in its structure is generally an amorphous solid. Examples of the amorphous solids are the glass and some kinds of plastic (Phillips, 1972; Falk and Langer, 1998; Angel et al., 1992, 2000). They are occasionally described as the supercooled liquids since their molecules are organized in an arbitrary as in the case of liquid state to some extent. For instance, glass is normally made from silicon dioxide (SiO_2) or the quartz sand, which possesses the crystalline structure. When this sand is melted and liquid is cooled quickly enough to avert the crystallization, glass which is an amorphous solid is formed (Hill, 1971; Scher and Montroll, 1975; Cahill and Pohl, 1987). Amorphous solids don't exhibit the sharp phase change from the solid to liquid at the definite melting point but instead soften slowly when these solids are heated. The physical properties of the amorphous solids are equal in all of the directions along any axis thus they are said to possess isotropic properties (Paliwal et al., 1994; Alexander, 1998; Hagrman et al., 2001) (Figure 6.3).

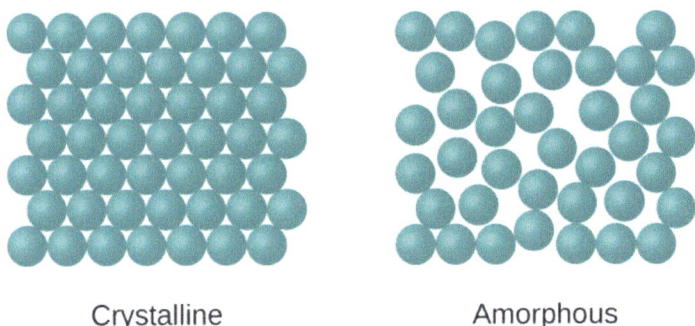

Crystalline   Amorphous

Figure 6.3. Crystal structure of the amorphous and crystalline solids.

Source: *https://opentextbc.ca/chemistry/chapter/10-5-the-solid-state-of-matter/.*

6.3.2. Crystalline Solids

More than 90% of the naturally existing and artificially made solids are crystalline. Minerals, clay, sand, limestone, carbon, metals, salts, all has the crystalline structures. A crystal is a regular, repeating arrangement of molecules or atoms (Alexander and Animalu, 1977; Schmidt, 1985; Vippagunta et al., 2001). The majority of the solids, including all of the metals, adopt the crystalline arrangement since the amount of stabilization accomplished by attaching interactions between adjacent particles larger when the particles form regular arrangements. In crystalline arrangement, the particles efficiently pack together to reduce the total intermolecular energy (Sherby and Burke, 1968; King-Smith and Vanderbilt, 1993).

The regular repeating configuration that the atoms assemble in is known as the crystalline lattice. The STM (scanning tunneling microscope) makes it probable to duplicate the electron cloud linked individual atoms at the surface of the material. Below is a STM image of the platinum surface exhibiting the regular arrangement of atoms (Kröger and Vink, 1956; Coble, 1961; Phillips, 1979) (Figure 6.4).

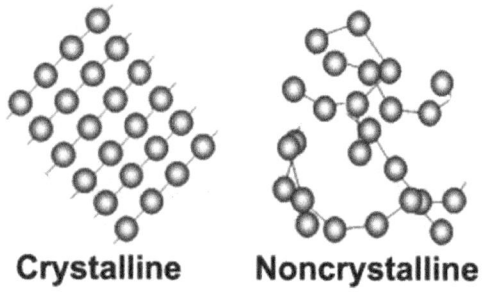

Crystalline   Noncrystalline

Figure 6.4. Difference between non-crystalline and crystalline structure.

Source: https://geologycafe.com/class/chapter3.html.

6.4. CRYSTAL STRUCTURE

Crystal structures might be suitably specified by relating the organization within the solid of a small representative group of molecules or atoms, known as the 'unit cell.' By multiplying the matching unit cells in 3 directions, the position of all particles is determined in the crystal. In nature, 14 distinct kinds of crystal lattices or structures are found (Ibers and Hamilton, 1964; Betteridge et al., 2003). The easiest crystalline unit cell to image is the cubic, in which the atoms are aligned in a square, three-dimensional grid. The unit cell is a box having an atom at every corner. Simple cubic crystals are comparatively rare, mostly since they incline to easily distort. However, numerous crystals form BCC (body-centered-cubic) or FCC (face-centered-cubic) structures, which are generally cubic with an additional atom centered in cube or centered in every face of the cube (Kay et al., 1964; Sheldrick, 2015). Most metals form FCC, BCC or HCP (hexagonal close packed) structures; though, the structure can vary depending on the temperature.

Crystalline structure is significant since it donates to properties of the material. For instance, it is simpler for planes of the atoms to slide by one another if those specific planes are carefully packed. Thus, lattice structures with strictly packed planes allow more plastic deformation as compared to those that are not closely packed. In addition, cubic lattice structures permit slippage to take place more easily as compared to the non-cubic lattices (Palczewski et al., 2000; Spek, 2003). This is since their symmetry offers closely packed planes in various directions. The FCC crystal structure will display more ductility than the bcc structure. The bcc lattice, even though cubic, isn't packed closely and creates strong metals. Tungsten and alpha-iron have the BCC form. The FCC lattice is both cubic and strictly packed and creates more ductile materials. Gamma-iron, silver, lead, and gold have FCC structures. Finally, HCP lattices are packed closely but aren't cubic. HCP metals like zinc and cobalt aren't as ductile as FCC metals (Olson et al., 1981; Gelato and Parthé, 1987).

As already mentioned, there are 14 distinct kinds of crystal unit cell lattices or structures found in nature. However, most of the metals and numerous other solids possess unit cell structures defined as bcc (body center cubic), FCC (face-centered cubic) or HCP (Figure 6.5).

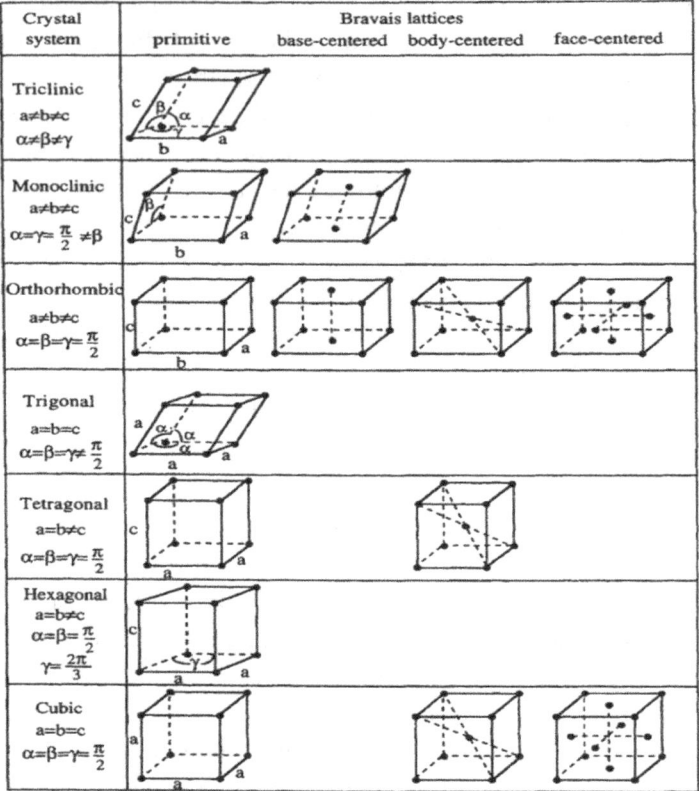

Figure 6.5. 14 kinds of crystal systems.

Source: https://www.quora.com/Which-crystal-systems-do-not-have-a-body-centered-lattice.

6.4.1. BCC (Body-Centered Cubic) Structure

The bcc unit cell has the atoms at every eight corners of the cube plus one atom in the center of cube. Every corner atom in the corner of the other cube so these corner atoms are mutual amongst the eight unit cells. It is described to have the coordination number of eight (Parrinello and Rahman, 1980). The body-centered cubic unit cell comprises of net total of two atoms; one in center and eight eighths from the corners atoms as exhibited in middle image below (middle image below). The image below outlines the unit cell in larger portion of the lattice (Figure 6.6).

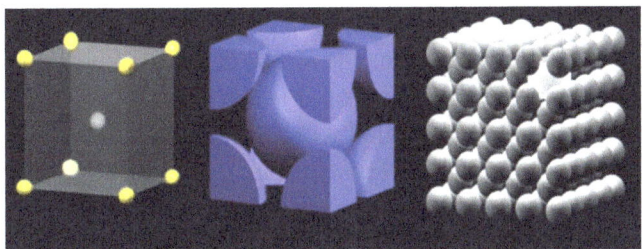

Figure 6.6. Body centered-cubic crystal structure.

Source: https://www.nde-ed.org/EducationResources/CommunityCollege/Materials/Structure/metallic_structures.htm.

The body centered-cubic arrangement does not allow the atoms to pack as strictly as the HCP or FCC arrangements. The bcc structure is frequently the high-temperature shape of metals that are closely-packed at the lower temperatures. The volume of the atoms in the cell per total volume of the cell is known as the packing factor. The body-centered cubic unit cell has a packing factor of 0.68.

6.4.2. FCC (Face Centered Cubic) Structure

The FCC structure has atoms positioned at every corner and at the centers of all cubic faces. Every corner atom in the corner of the other cube so that the corner atoms are mutual amongst the eight-unit cells. In addition, each of its 6 face-centered atoms in common with the neighboring atom. As 12 of the atoms are mutual, it is considered to have the coordination number of 12. The FCC unit cell comprises of net total of 4 atoms; 8 eighths from the corners atoms and 6 halves of face atoms as exhibited in middle image below. The image below outlines the unit cell in larger portion of the lattice (Figure 6.7).

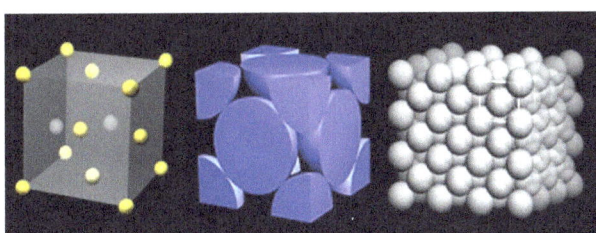

Figure 6.7. Face centered-cubic structure of atoms.

Source: https://www.nde-ed.org/EducationResources/CommunityCollege/Materials/Structure/metallic_structures.htm.

In the FCC structure, the atoms can pack nearer together as compared to the body-centered cubic structure. The atoms of the one layer nest themselves in vacant space amongst the atoms of adjacent layer. To picture the packing arrangement, imagine the box filled with layer of balls which are positioned in rows and columns. When the few extra balls are tossed inbox, they won't balance on top of balls directly in the first layer but rather will come to the rest in pocket created amongst four balls of the bottom layer. The packing factor is 0.74 for the FCC crystals. Some metals having the FCC structure comprise aluminum, gold, copper, iridium, lead, silver, platinum, and nickel.

6.4.3. HCP (Hexagonal Close Packed) Structure

Another usual close-packed structure is HCP. The hexagonal structure of interchanging layers is moved so its atoms are positioned to the gaps of preceding layer. The atoms of one layer nest themselves in vacant empty space amongst the atoms of near layer such as in the FCC structure. However, rather than being the cubic structure, the pattern is usually hexagonal (Figure 6.8).

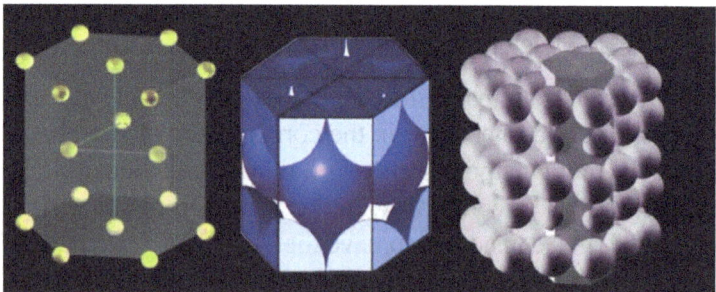

Figure 6.8. Hexagonal close packed structure of the atoms.

Source: https://www.nde-ed.org/EducationResources/CommunityCollege/Materials/Structure/metallic_structures.htm.

The hexagonal close-packed structure has three layers of the atoms. In each of the bottom and top layers, there are six atoms that organize themselves in shape of the hexagon and the seventh atom sits in middle of the hexagon. The middle layer possesses three atoms nestle in triangular grooves of the bottom and top plane. Observe that there are six of these grooves surrounding every atom in the hexagonal plane but only three of the grooves can be filled by the atoms.

As exhibited in the middle image above, six atoms are present in the hexagonal close-packed unit cell. Each of the twelve atoms in corners of the bottom and top layers donates 1/6 atom to unit cell, the 2 atoms in center of hexagon of both the bottom and top layers each donate ½ atom and each of 3 atoms in middle layer donate one atom. The image on right above tries to display various HCP unit cells in the larger lattice.

The coordination number of atoms in this particular structure is 12. There are six adjacent neighbors in the same strictly packed layer, three in the layer below and three in the layer above. The packing factor is around 0.74, which is similar to the FCC unit cell.

6.5. SOLIDIFICATION

The crystallization of a large amount of the material from the single point of nucleation results in a single crystal. In the engineering materials, the single crystals are made only under cautiously controlled conditions (Flemings, 1974; Rohatgi et al., 1986). The expense of creating single-crystal materials is justified for exceptional applications, like the turbine engine blades, piezoelectric materials, and solar cells. Usually, when the material starts to solidify, several crystals start to grow in liquid and the polycrystalline solid forms.

The moment crystal starts to grow is called the nucleation and point where it takes place is the nucleation point. At solidification temperature, the atoms of the liquid, like the melted metal, start to bond together at nucleation points and begin to form crystals. The final sizes of individual crystals are dependent on number of the nucleation points. The crystals grow in size by progressive addition of the atoms and develop until they impose upon neighboring growing crystal (Hunt and Jackson, 1966; Chalmers, 1970) (Figure 6.9).

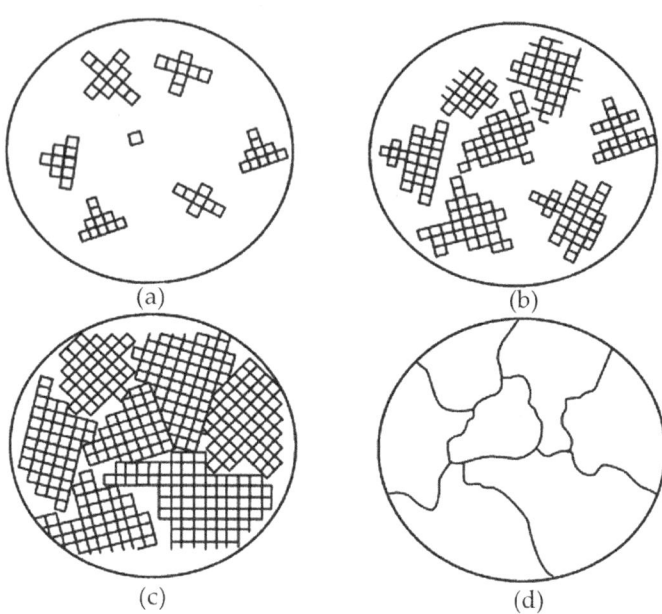

Figure 6.9. (a) Nucleation of crystals, (b) crystal development, (c) irregular grains develop as the crystals grow together, (d) grain boundaries as viewed in the microscope.

Source: https://www.nde-ed.org/EducationResources/CommunityCollege/Materials/Structure/solidification.htm.

In the engineering materials, crystal is normally referred to as the grain. A grain is simply the crystal without smooth faces since its growth was obstructed by interaction with another grain or the boundary surface. The interface made between the grains is known as the grain boundary. The atoms amongst the grains have no crystalline structure and are disordered (Kurz et al., 1986; Boettinger et al., 2002).

Grains are occasionally large enough to be observable under the usual light microscope or viewed even to the naked eye. The spangles which are viewed on the newly galvanized metals are the grains. Quick cooling generally outcomes in more nucleation points and the smaller grains. Slow cooling generally outcomes in the larger grains which will possess lower strength, ductility, and hardness.

6.5.1. Dendrites

In the metals, the crystals which are formed in liquid during freezing usually follow the pattern comprising of the main branch with numerous appendages. The crystal having this morphology resembles the pine tree to some extent and is known as a dendrite, which means branches. The creation of dendrites takes place as crystals grow in the defined planes because of the crystal lattice they form. The figure in the right displays how the cubic crystal can progress in the melt in 3 D, which resembles the 6 faces of the cube. For clarity of drawing, adding of the unit cells with constant solidification from 6 faces is exhibited simply as lines. The Secondary dendrite arms branch off primary arm, and the tertiary arms off etcetera and the secondary arms (Figure 6.10).

Figure 6.10. Dendritic structure because of solidification.

Source: https://www.doitpoms.ac.uk/tlplib/solidification_alloys/dendritic.php.

During freezing of the polycrystalline material, numerous dendritic crystals are formed and grow till they ultimately become large enough to intrude upon each other. Finally, the interdendritic spaces amongst the dendrite arms crystallize to produce the more regular crystal. The original dendritic pattern might not be obvious when inspecting the microstructure of the material. However, dendrites can frequently be observed in solidification spaces that sometimes happen in castings or the welds.

6.5.2. Shrinkage

Most materials shrink or contract during cooling and solidification. Shrinkage is an outcome of:

- Contraction of liquid as it cools down before its solidification;
- Contraction during the phase change from the liquid to solid;
- Contraction of solid cooling the surroundings.

Shrinkage at times can cause cracking to take place in the component as it solidifies. As the coolest area of the volume of liquid is, one where it contacts the die or mold, solidification normally starts first at this surface. As the crystals develop inward, the material carries on to shrink. If the solid surface is quite rigid and won't deform to lodge the internal shrinkage, the stresses might become high to surpass the tensile strength of material and cause the crack to form. Shrinkage cavitation sometimes takes place because as the material solidifies inward, the shrinkage happened to such extent that there are not enough atoms available to fill the vacant space and a gap is left (Figure 6.11).

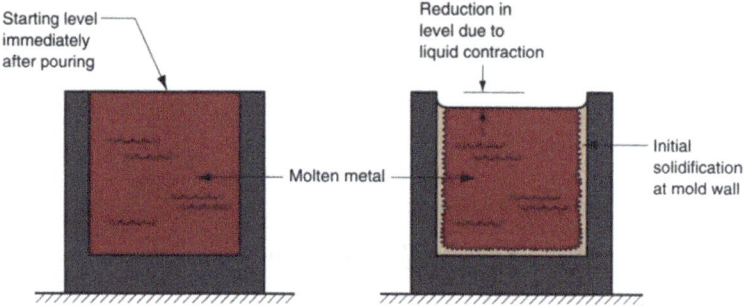

Figure 6.11. Shrinkage because of solidification of the liquid metal.

Source: https://slideplayer.com/slide/4338191/.

6.6. ANISOTROPY AND ISOTROPY

In the single crystal, the mechanical and physical properties frequently differ with orientation. It can be viewed from looking at models of the crystalline structure that atoms must be capable to slip over each other or change in relation to each other easier in some orientations than others (Kinoshita and Mura, 1971; He, 1990; Blatte et al., 1992). When properties of the material change with different crystallographic directions, the material is known to be anisotropic (Berryman, 1979; Barrow and Tsagas, 2005; Shankar and Rayappan, 2016).

Consecutively, when the properties of the material are same in all the directions, the material is known to be isotropic. For numerous polycrystalline materials, the grain directions are arbitrary before any work is exerted on the material (Starobinskii, 1985; Crawley and Lazarus, 1991). Thus, even if individual grains are said to be anisotropic, the property dissimilarities incline to average out and overall, the material is isotropic. When the material is made, the grains are normally distorted and extended in one or more than one direction which tends to make the material anisotropic (Beaulieu and Allen, 1994; Cances et al., 1997).

6.7. CRYSTAL DEFECTS

The perfect ideal crystal, with atoms of the same kind in the right position is a matter of theoretical modeling. All crystals do have some defects. Defects contribute to the mechanical properties of the metals. Using the term defect is sort of misnomer as these features are normally intentionally used to control the mechanical properties of the material (Stukowski and Albe, 2010; Bollmann, 2012). Adding alloying elements to the metal is one method of introducing the crystal defect. Nevertheless, the term defect will be utilized, just remember the point that crystalline defects aren't always bad. There are fundamental classes of the crystal defects (Burt and Mitchell, 1981; Ludwig et al., 2001):

- Point defects, which are the places or positions where an atom is absent or irregularly located in the lattice structure. Point defects comprise lattice vacancies, substitution impurity atoms, self-interstitial atoms, and interstitial impurity atoms.
- Linear defects, which are the groups of atoms in the irregular positions. Linear defects are usually called dislocations.
- Planar defects, which are the interfaces amongst homogeneous regions of material. Planar defects comprise grain boundaries, external surfaces, and stacking faults (Figure 6.12).

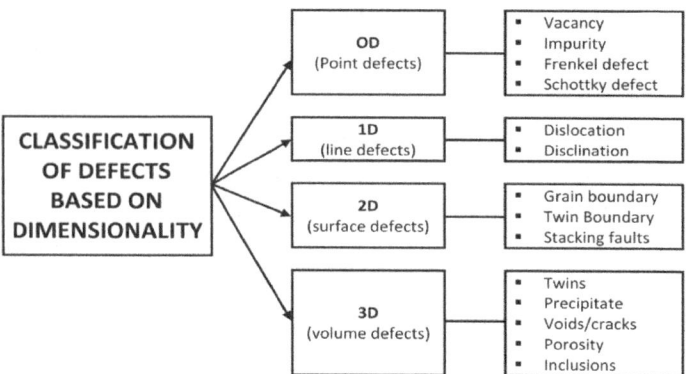

Figure 6.12. Different kinds of defects existent in crystals.

Source: https://iopscience.iop.org/book/978-1-6817-4473-5/chapter/bk978-1-6817-4473-5ch4.

It is vital to observe at this point that the plastic deformation in the material occurs because of the movement of dislocations. Millions of dislocations outcome for the plastic forming operations like rolling and extruding. It is also significant to observe that any defect in regular lattice structure disturbs the motion of the dislocation, which makes plastic deformation or slip more difficult. These defects not just include the planer and point defects described above but also other dislocations. Dislocation movement creates additional dislocations and when these dislocations run into one another, it frequently impedes the movement of dislocations. This pushes up the force required to transfer the dislocation or strengthens the material (Casey et al., 1988).

REFERENCES

1.   Alexander, O. E., & Animalu, E., (1977). *Intermediate Quantum Theory of Crystalline Solids* (Vol. 1, pp. 1–18). Englewood Cliffs: Prentice-Hall Inc.

2.   Alexander, S., (1998). Amorphous solids: Their structure, lattice dynamics, and elasticity. *Physics Reports, 296*(2–4), 65–236.

3.   Angell, C. A., (1992). Mobile ions in amorphous solids. *Annual Review of Physical Chemistry, 43*(1), 693–717.

4.   Angell, C. A., Ngai, K. L., McKenna, G. B., McMillan, P. F., & Martin, S. W., (2000). Relaxation in glass forming liquids and amorphous solids. *Journal of Applied Physics, 88*(6), 3113–3157.

5.   Barrow, J. D., & Tsagas, C. G., (2005). New isotropic and anisotropic sudden singularities. *Classical and Quantum Gravity, 22*(9), 1563.

6.   Beaulieu, C., & Allen, P. S., (1994). Determinants of anisotropic water diffusion in nerves. *Magnetic Resonance in Medicine, 31*(4), 394–400.

7.   Berryman, J. G., (1979). Long-wave elastic anisotropy in transversely isotropic media. *Geophysics, 44*(5), 896–917.

8.   Betteridge, P. W., Carruthers, J. R., Prout, K., & Watkin, D. J., (2003). CRYSTALS version 12: Software for guided crystal structure analysis. *Journal of Applied Crystallography, 36*(6), 1–22.

9.   Blatter, G., Geshkenbein, V. B., & Larkin, A. I., (1992). From isotropic to anisotropic superconductors: A scaling approach. *Physical Review Letters, 68*(6), 875.

10. Boettinger, W. J., Warren, J. A., Beckermann, C., & Karma, A., (2002). Phase-field simulation of solidification. *Annual Review of Materials Research, 32*(1), 163–194.

11. Bollmann, W., (2012). *Crystal Defects and Crystalline Interfaces* (Vol. 1, pp. 16–20). Springer Science & Business Media.

12. Burt, H. M., & Mitchell, A. G., (1981). Crystal defects and dissolution. *International Journal of Pharmaceutics, 9*(2), 137–152.

13. Cahill, D. G., & Pohl, R. O., (1987). Thermal conductivity of amorphous solids above the plateau. *Physical Review B, 35*(8), 4067.

14. Cances, E., Mennucci, B., & Tomasi, J., (1997). A new integral equation formalism for the polarizable continuum model: Theoretical background and applications to isotropic and anisotropic dielectrics. *The Journal of Chemical Physics, 107*(8), 3032–3041.

15. Casey, W. H., Carr, M. J., & Graham, R. A., (1988). Crystal defects and the dissolution kinetics of rutile. *Geochimica et Cosmochimica Acta, 52*(6), 1545–1556.

16. Chalmers, B., (1970). Principles of solidification. In: *Applied Solid-State Physics* (Vol. 1, pp. 161–170). Springer, Boston, MA.

17. Coble, R. L., (1961). Sintering crystalline solids. I. Intermediate and final state diffusion models. *Journal of Applied Physics, 32*(5), 787–792.

18. Crawley, E. F., & Lazarus, K. B., (1991). Induced strain actuation of isotropic and anisotropic plates. *AIAA Journal, 29*(6), 944–951.

19. De Graef, M., & McHenry, M. E., (2012). *Structure of Materials: An Introduction to Crystallography, Diffraction and Symmetry* (Vol. 1, pp. 1–16). Cambridge University Press.

20. Dmitriev, S. V., Li, J., Yoshikawa, N., Tanaka, Y., Kagawa, Y., Kitamura, T., & Yip, S., (2004). Breaking atomic bonds through vibrational mode localization. In: *Defect and Diffusion Forum* (Vol. 233, pp. 49–60). Trans Tech Publications.

21. Ehrenstein, G. W., (2012). *Polymeric Materials: Structure, Properties, Applications* (Vol. 1, pp. 1–10). Carl Hanser Verlag GmbH Co KG.

22. Evers, W. H., Schins, J. M., Aerts, M., Kulkarni, A., Capiod, P., Berthe, M., & Peters, J. L., (2015). High charge mobility in two-dimensional percolative networks of PbSe quantum dots connected by atomic bonds. *Nature Communications, 6*, 8195.

23. Falk, M. L., & Langer, J. S., (1998). Dynamics of viscoplastic deformation in amorphous solids. *Physical Review E, 57*(6), 7192.

24. Flemings, M. C., (1974). Solidification processing. *Metallurgical Transactions, 5*(10), 2121–2134.

25. Fratzl, P., (2008). Collagen: Structure and mechanics, an introduction. In: *Collagen* (Vol. 1, pp. 1–13). Springer, Boston, MA.

26. Gelato, L. M., & Parthé, E., (1987). Structure tidy – a computer program to standardize crystal structure data. *Journal of Applied Crystallography, 20*(2), 139–143.

27. Hagrman, P. J., Finn, R. C., & Zubieta, J., (2001). Molecular manipulation of solid-state structure: Influences of organic components on vanadium oxide architectures. *Solid State Sciences, 3*(7), 745–774.

28. Haubold, T., Birringer, R., Lengeler, B., & Gleiter, H., (1989). EXAFS studies of nanocrystalline materials exhibiting a new solid-state

structure with randomly arranged atoms. *Physics Letters A, 135*(8–9), 461–466.

29. He, X. F., (1990). Anisotropy and isotropy: A model of fraction-dimensional space. *Solid State Communications, 75*(2), 111–114.

30. Hill, R. M., (1971). Poole-Frenkel conduction in amorphous solids. *Philosophical Magazine, 23*(181), 59–86.

31. Hunt, J. D., & Jackson, K. A., (1966). Binary eutectic solidification. *Trans. Metall. Soc. AIME, 236*(6), 843–852.

32. Hüsing, N., & Schubert, U., (1998). Aerogels—airy materials: Chemistry, structure, and properties. *Angewandte Chemie International Edition, 37*(1/2), 22–45.

33. Ibers, J. A., & Hamilton, W. C., (1964). Dispersion corrections and crystal structure refinements. *Acta Crystallographica, 17*(6), 781–782.

34. Johnston, A. G., Leigh, D. A., Pritchard, R. J., & Deegan, M. D., (1995). Facile synthesis and solid-state structure of a benzylic amide [2] Catenane. *Angewandte Chemie International Edition in English, 34*(11), 1209–1212.

35. Jutzi, P., Becker, A., Stammler, H. G., & Neumann, B., (1991). Synthesis and solid-state structure of (Me3Si) 3CGeCH (SiMe3) 2, a monomeric dialkyl germylene. *Organometallics, 10*(6), 1647–1648.

36. Kay, M. I., Young, R. A., & Posner, A. S., (1964). Crystal structure of hydroxyapatite. *Nature, 204*(4963), 1050–1052.

37. Kimura, C., Sota, H., Aoki, H., & Sugino, T., (2009). Atomic bonds in boron carbon nitride films synthesized by remote plasma-assisted chemical vapor deposition. *Diamond and Related Materials, 18*(2/3), 478–481.

38. King-Smith, R. D., & Vanderbilt, D., (1993). Theory of polarization of crystalline solids. *Physical Review B, 47*(3), 1651.

39. Kinoshita, N., & Mura, T., (1971). Elastic fields of inclusions in anisotropic media. *Physica. Status Solidi(a), 5*(3), 759–768.

40. Klok, H. A., & Lecommandoux, S., (2006). Solid-state structure, organization, and properties of peptide—synthetic hybrid block copolymers. In: *Peptide Hybrid Polymers* (Vol. 1, pp. 75–111). Springer, Berlin, Heidelberg.

41. Kröger, F. A., & Vink, H. J., (1956). Relations between the concentrations of imperfections in crystalline solids. In: *Solid State Physics* (Vol. 3, pp. 307–435). Academic Press.

42. Kurz, W. l., Giovanola, B., & Trivedi, R., (1986). Theory of microstructural development during rapid solidification. *Acta Metallurgica., 34*(5), 823–830.

43. Larson, R. G., (1999). *The Structure and Rheology of Complex Fluids* (Vol. 150, pp. 1–18). New York: Oxford university press.

44. Lawniczak-Jablonska, K., Suski, T., Gorczyca, I., Christensen, N. E., Libera, J., Kachniarz, J., & Grzegory, I., (2002). Anisotropy of atomic bonds formed by p-type dopants in bulk GaN crystals. *Applied Physics A, 75*(5), 577–583.

45. Ludwig, W., Cloetens, P., Härtwig, J., Baruchel, J., Hamelin, B., & Bastie, P., (2001). Three-dimensional imaging of crystal defects bytopo-tomography'. *Journal of Applied Crystallography, 34*(5), 602–607.

46. Mas-Torrent, M., & Rovira, C., (2011). Role of molecular order and solid-state structure in organic field-effect transistors. *Chemical Reviews, 111*(8), 4833–4856.

47. McKeown, N. B., (1998). *Phthalocyanine Materials: Synthesis, Structure and Function* (Vol. 6, pp. 1–30). Cambridge University Press.

48. Megaw, H. D., (1958). Some empirical problems concerning atomic bonds. *Reviews of Modern Physics, 30*(1), 96.

49. Meyers, M. A., Chen, P. Y., Lin, A. Y. M., & Seki, Y., (2008). Biological materials: Structure and mechanical properties. *Progress in Materials Science, 53*(1), 1–206.

50. Newnham, R. E., (2005). *Properties of Materials: Anisotropy, Symmetry, Structure* (Vol. 1, pp. 1–16). Oxford University Press on Demand.

51. Olson, D. H., Kokotailo, G. T., Lawton, S. L., & Meier, W. M., (1981). Crystal structure and structure-related properties of ZSM-5. *The Journal of Physical Chemistry, 85*(15), 2238–2243.

52. Ono, H., & Koyanagi, K. I., (2000). Infrared absorption peak due to Ta = O bonds in Ta 2 O 5 thin films. *Applied Physics Letters, 77*(10), 1431–1433.

53. Palczewski, K., Kumasaka, T., Hori, T., Behnke, C. A., Motoshima, H., Fox, B. A., & Yamamoto, M., (2000). Crystal structure of rhodopsin: AG protein-coupled receptor. *Science, 289*(5480), 739–745.

54. Paliwal, S., Geib, S., & Wilcox, C. S., (1994). Molecular torsion balance for weak molecular recognition forces. Effects of" tilted-T" edge-to-

face aromatic interactions on conformational selection and solid-state structure. *Journal of the American Chemical Society, 116*(10), 4497–4498.

55. Parrinello, M., & Rahman, A., (1980). Crystal structure and pair potentials: A molecular-dynamics study. *Physical Review Letters, 45*(14), 1196.

56. Phillips, J. C., (1979). Topology of covalent non-crystalline solids I: Short-range order in chalcogenide alloys. *Journal of Non-Crystalline Solids, 34*(2), 153–181.

57. Phillips, W. A., (1972). Tunneling states in amorphous solids. *Journal of Low Temperature Physics, 7*(3/4), 351–360.

58. Prashanth, K. H., Kittur, F. S., & Tharanathan, R. N., (2002). Solid state structure of chitosan prepared under different N-deacetylating conditions. *Carbohydrate Polymers, 50*(1), 27–33.

59. Rohatgi, P. K., Asthana, R., & Das, S., (1986). Solidification, structures, and properties of cast metal-ceramic particle composites. *International Metals Reviews, 31*(1), 115–139.

60. Scher, H., & Montroll, E. W., (1975). Anomalous transit-time dispersion in amorphous solids. *Physical Review B, 12*(6), 2455.

61. Schmidt, H., (1985). New type of non-crystalline solids between inorganic and organic materials. *Journal of Non-Crystalline Solids, 73*(1–3), 681–691.

62. Shankar, P., & Rayappan, J. B. B., (2016). Racetrack effect on the dissimilar sensing response of ZnO thin film: An anisotropy of isotropy. *ACS Applied Materials and Interfaces, 8*(37), 24924–24932.

63. Sheldrick, G. M., (2015). Crystal structure refinement with SHELXL. *Acta Crystallographica Section C: Structural Chemistry, 71*(1), 3–8.

64. Sherby, O. D., & Burke, P. M., (1968). Mechanical behavior of crystalline solids at elevated temperature. *Progress in Materials Science, 13*, 323–390.

65. Sheth, J. P., Xu, J., & Wilkes, G. L., (2003). Solid state structure—property behavior of semi crystalline poly (ether-block-amide) PEBAX® thermoplastic elastomers. *Polymer, 44*(3), 743–756.

66. Smith, W. F., Hashemi, J., & Wang, S. H., (2006). *Foundations of Materials Science and Engineering* (Vol. 397, pp. 1–40). New York: McGraw-hill.

67. Spek, A. L. J., (2003). Single-crystal structure validation with the program PLATON. *Journal of Applied Crystallography, 36*(1), 7–13.

68. Starobinskii, A. A., (1985). Cosmic background anisotropy induced by isotropic flat-spectrum gravitational-wave perturbations. *Soviet Astronomy Letters, 11*, 133–136.

69. Stukowski, A., & Albe, K., (2010). Extracting dislocations and non-dislocation crystal defects from atomistic simulation data. *Modeling and Simulation in Materials Science and Engineering, 18*(8), 085001.

70. Sutton, A. P., (1993). *Electronic Structure of Materials* (Vol. 1, pp. 1–10). Clarendon Press.

71. Tanaka, K., Inui, H., Yamaguchi, M., & Koiwa, M., (1999). Directional atomic bonds in $MoSi_2$ and other transition-metal disilicides with the C11b, C40 and C54 structures. *Materials Science and Engineering: A, 261*(1/2), 158–164.

72. Vippagunta, S. R., Brittain, H. G., & Grant, D. J., (2001). Crystalline solids. *Advanced Drug Delivery Reviews, 48*(1), 3–26.

73. Volokh, K. Y., (2008). Multiscale modeling of material failure: From atomic bonds to elasticity with energy limiters. *International Journal for Multiscale Computational Engineering, 6*(5), 1–11.

74. Welker, J., & Giessibl, F. J., (2012). Revealing the angular symmetry of chemical bonds by atomic force microscopy. *Science, 336*(6080), 444–449.

75. Wong, W. S., & Salleo, A., (2009). *Flexible Electronics: Materials and Applications* (Vol. 11, pp. 5–33). Springer Science & Business Media.

76. Wrinch, D., & Harker, D., (1940). Lengths and strengths of atomic bonds. *The Journal of Chemical Physics, 8*(6), 502–503.

77. Xia, J., Bacon, J. W., & Jasti, R., (2012). Gram-scale synthesis and crystal structures of [8]-and [10] CPP, and the solid-state structure of C 60@[10] CPP. *Chemical Science, 3*(10), 3018–3021.

Chapter

7

Mechanical Properties and Performance of Materials

CONTENTS

7.1. INTRODUCTION

Samples of the engineering materials are exposed to a broad variety of mechanical tests to measure their elastic constants, strength, and other material properties along with their performance under the variety of actual utilization conditions and environments (Curtis and Clark, 1990; Oka and Yoshida, 2005). The outcomes of these tests are used for two main purposes: 1) engineering design and 2) the quality control either by materials producer to validate the procedure or by the end-user to endorse the material specifications (Kolsky, 1949; Murayama, 1978).

Due to the necessity to compare the performance and measured properties on the common basis, producers, and users of the materials use standardized test techniques like those developed by ASTM (American Society for Testing and Materials) and ISO (International Organization for Standardization). ISO and ASTM are two of several standards-writing professional organizations in the world (Barker and Hollenbach, 1965; Armstrong, 1970). These standards recommend the technique by which a test sample will be made and tested, along with how the test outcomes will be examined and reported. Standards are also present which define the terminology and nomenclature along with classification and the specification schemes (Hudson, Liu, and Crampin, 1996; Fundenberger et al., 1997).

7.2. MECHANICAL BEHAVIOR

For the structural material, the questions can be asked: "How strong the material is?" "How much deformation will the material undergo?" The answers to these questions decide the mechanical behavior of the material. Mechanical behavior describes the material's response to the load (Mecklenburg et al., 2012; Cheng, Jiang, and Tang, 2014).

Metals are most generally linked with structural applications; though engineering polymers and ceramics are also used in structural applications. To get a grasp on the material's behavior, the material properties are well-defined by the professional organizations like ASTM, ASM, ANSI, etc. (Duncan and Chang, 1970; Yang et al., 2002; Ilie et al., 2013). The properties are demarcated according to sensibly designed lab tests that try to duplicate as nearly as feasible the service conditions which the material will encounter (Gupta and Wong, 2005; Cheng et al., 2009). These material properties might depend on temperature, UV radiation, moisture, or other factors, even the atmospheric hydrogen or oxygen (Sugimura et al., 1997; Westman, 2001; Jayaraman and Bhattacharyya, 2004).

Structural engineers decide stresses and the stress distributions that will be faced by the materials for different loading conditions. It is the work of a material engineer to find out how to develop and fabricate the materials that will bear these stresses (Barthelat et al., 2006; Monteiro et al., 2008; Neal et al., 2015).

7.3. STRESS AND STRAIN

7.3.1. Stress

The term stress is employed to describe loading in terms of the force applied to a particular cross-sectional area of the object (Berthelot and Ling, 1999; Dowling, 2012). From the viewpoint of loading, stress(s) is the system of forces or applied force that tends to deform the body. From the viewpoint of what is taking place within the material, stress (s) is the internal distribution of the forces within the body that balance and react to loads applied to the body. The stress distribution may not or may be uniform, depending on the nature of the loading conditions (Haynes, 1981; Sánchez-Arévalo and Pulos, 2008). For instance, bar loaded in pure tension will fundamentally have a uniform tensile stress distribution. On the other hand, a bar loaded in the bending will have the stress distribution that varies with the distance perpendicular to a normal axis (Sherby and Burke, 1968; Guo, 2003).

Simplifying assumptions are frequently utilized to represent stress as the vector quantity for several engineering calculations and material property determination (Fink and Carlsen, 1978; Van Dick and Wagner, 2001). The word *vector* normally refers to the quantity possesses magnitude and direction. For instance, the stress in the axially loaded bar is equal to the force applied divided by the cross-sectional area of the bar (Hutchinson, 1968; Lade and Duncan, 1975) (Figure 7.1).

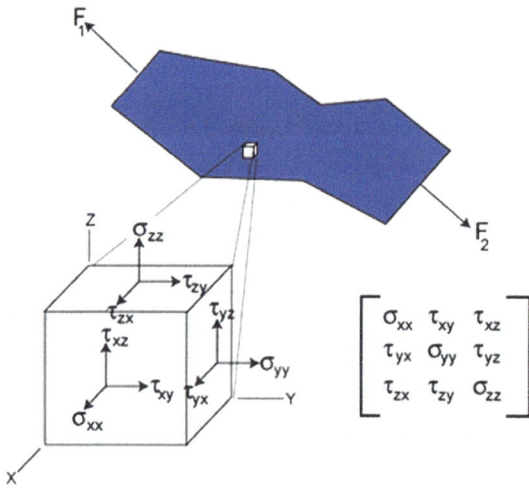

$$\text{Stress, } \sigma = \frac{\text{Force}}{\text{Cross-Sectional Area}} = \frac{F}{A_o}$$

Figure 7.1. Diagram of forces acting on the rod to produce stress.

Source: https://www.nde-ed.org/EducationResources/CommunityCollege/Materials/Mechanical/StressStrain.htm.

Some common measurement units of the stress are (Davies and Tripathi, 1993; Altman et al., 2002) (Figure 7.2):

Psi = pounds per square inch (lbs/in$^{2)}$

ksi or kpsi = 1000 pounds per square inch (kilopounds/in^2)

Pa = Newtons per square meter or pascals (N/m^2)

kPa = 1000 Newtons per square meter (Kilopascals)

GPa = (1 million Newtons per square meter (Gigapascals)

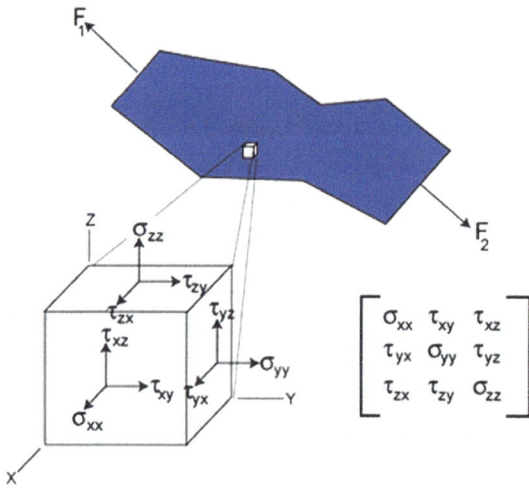

Figure 7.2. Components of the stress in the 3-dimensional object.

Source: https://www.nde-ed.org/EducationResources/CommunityCollege/Materials/Mechanical/StressStrain.htm.

It must be observed that the stresses in two-Dimensional or three-Dimensional solids are essentially more complex and must be stated more methodically (Bassett and Becker, 1962; Titze, 1994). The internal force which is acting on the small area of the plane can be distributed into three components: one normal to plane and the two parallel to plane. The component of normal force divided by area provides the normal stress, and the components of parallel force divided by area provide the shear stress (De Wolf, 1996; Chiquet, 1999). These stresses are the average stresses as the area is finite; however, when the area is allowed to approach zero, these stresses become the stresses at a point. As stresses are described in association to the plane which passes through the point under attention and number of such kind of planes is infinite, there seem an infinite set of the stresses at the point (Sadoshima and Izumo, 1997; Nomura and Takano-Yamamoto, 2000). By chance, it can be demonstrated that stresses on any plane can also be figured from stresses on the three orthogonal planes advancing through the point. As every plane has three stresses, stress tensor has nine stress components, which define the state of the stress at the point (Komuro and Yazaki, 1993; Xu et al., 1999).

7.3.2. Strain

It is a response of the system to the stress applied. When the material is loaded with the force, it creates stress, which causes the material to deform. Engineering strain is well-defined as the amount of the deformation in direction of the force applied divided by the preliminary length of the material (Rubin and Lanyon, 1985; Duncan and Turner, 1995). This outcome in the unitless number, even though it is frequently left in non-simple forms, like meters per meter or inches per inch. For instance, the strain in the bar that is being overextended in tension is the amount of the elongation or variation in length divided by the original length. As in the situation of the stress, the distribution of strain may not or may be uniform in the complex structural element, which is dependent on the nature of the loading condition (Kim et al., 1999; Ehrlich and Lanyon, 2002) (Figure 7.3).

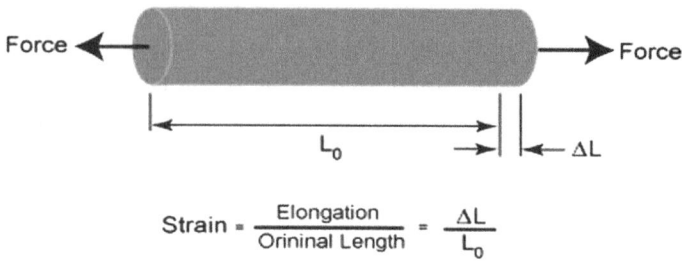

$$\text{Strain} = \frac{\text{Elongation}}{\text{Orininal Length}} = \frac{\Delta L}{L_0}$$

Figure 7.3. Diagram of strain developed in the rod.

Source: https://www.nde-ed.org/EducationResources/CommunityCollege/Materials/Mechanical/StressStrain.htm.

If stress is very small, the material might only strain the small amount and material will come back to its original size once the stress is removed. This is known as elastic deformation, like elastic, it comes back to the unstressed state (Jones et al., 1991; Owan et al., 1997; Elvin et al., 2001). Elastic deformation only takes place in the material when the stresses are lower as compared to critical stress known as the yield strength. If the material is loaded over its elastic limit, then the material will stay in the deformed condition once the load is removed. This is known as plastic deformation (Barlow et al., 1970; Lee et al., 2012).

7.3.3. Engineering and True Stress and Strain

The discussion above focused on the *engineering* stress and strain, which use fixed, undeformed cross-sectional area in calculations. True stress and strain measures deal with variations in the cross-sectional area by utilizing the instantaneous values for the area (Ling, 1996; Faridmehr et al., 2014). The stress-strain curve doesn't provide the true indication of deformation characteristics of the metal as it is entirely based on original dimensions of sample and these dimensions vary constantly during the testing utilized to produce the data (Kontou and Farasoglou, 1998; Khanafer et al., 2009).

Data on engineering stress and strain is commonly utilized since it is quite easier to produce the data and tensile properties are enough for the engineering calculations. When considering the curves of stress-strain in the next section, however, it must be understood that the metals and some other materials linger to strain-harden till they break and the stress needed to create further deformation also increases (Carreker Jr and Hibbard Jr, 1953; Joun et al., 2008).

- (Engineering or the nominal) (normal: $F \perp A \rightarrow$ T or C) stress = F/A_o.
- (Engineering or the nominal) (normal: T or C) strain = $\Delta l/l_o$.
- Elastic deformation $\rightarrow$ stretching of the atomic bonds.
- Plastic deformation $\rightarrow$distortion, breaking, and restoration of the atomic bonds.
- True strain, $\varepsilon = \ln \dfrac{l_i}{l_o}$ ($\varepsilon = \int_{lo}^{li} \dfrac{dl}{l}$).
- True stress, $\sigma = F/A_i$.
- Shear strain, $\gamma = \tan\theta$.
- Shear stress, $\tau = F/A_o$ where F // A.
- Hooke's law exhibits the linear association among stress and strain:

s = Ee for the normal stress

$\tau = G\gamma$ for the shear stress

For metals, Hooke's law relates to the elastic section where there are comparatively low values of the stress and strain (Zhu et al., 2015).

7.4. FUNDAMENTALS DEFINITIONS IN THE MECHANICS OF MATERIALS

7.4.1. Young's Modulus, E (Aka Modulus of Elasticity or Elastic Modulus)

E measures the confrontation to the separation of the neighboring atoms, i.e., the interatomic bonding forces (Black, 2005; Vegas and Del Yerro, 2013). Young's modulus is proportional to the slope of the atomic bonding force versus the atomic separation distance plot at an equilibrium point.

$$E \propto \frac{dF}{da}\Big|_{a=ao}$$

7.4.2. Tensile Modulus

Normally refers to Young's modulus for the stress-strain curves of the constant slope (Davim, 2008).

7.4.3. Elastic Modulus

Slope at the start of the curve if the slope of the stress-strain curve is not constant (Craig and Kurdila, 2006).

7.4.4. Tangent Modulus

The slope of the line tangent to the curve at POI (point of interest) if the slope of the stress-strain curve is not constant (Donald and Chen, 1997).

7.4.5. Secant Modulus

The slope of the line drawn from the origin of a curve to POI if the slope of the stress-strain curve is not constant (Hjelmstad, 2007).

7.4.6. Stiffness

It is stated as a property of the material which is firm and hard to bend (Oliver, 1996).

Stiffness $= E A_o / l_o$

7.4.7. Shear Modulus

It is defined as a ratio of the shear stress to the shear strain. It is also called the modulus of rigidity and is usually denoted by G.

7.4.8. Hardness

A simple substitute for a tensile test. The test measures confrontation to indentation by putting an indenter into a material with a load and thus calculating the hardness number. The hardness number is centered on a formula utilizing indentation geometry measurements. It is a non-destructive examination (NDE). Occasionally the hardness number can also be correlated to a tensile strength of the material.

7.5. TENSILE PROPERTIES

Tensile properties specify how the material will respond to forces applied in tension. The tensile test is an essential mechanical test where a cautiously prepared sample is loaded in a controlled manner whereas measuring the load applied and the elongation of a sample over some distance (Ascenzi and Bonucci, 1967; Drury et al., 2004). Tensile tests are used to define the elastic limit, elongation, modulus of elasticity, proportional limit, tensile strength, yield point, reduction in area, yield strength and the other tensile properties (Bigliani et al., 1992; Kelly and Tyson, 1965).

The key product of the tensile test is load versus the elongation curve which is converted into the stress vs. strain curve. As both the engineering strain and the engineering stress are attained by dividing the load and the elongation by the constant values, the load-elongation curve will normally have a similar shape as the shape of the engineering stress-strain curve (Skaggs et al., 1994; Shen et al., 2005). The stress-strain curve associates the stress applied to the subsequent strain and every material possesses its exclusive stress-strain curve. A usual engineering stress-strain curve is displayed below (Dalla Torre et al., 2002; Ku et al., 2011) (Figure 7.4).

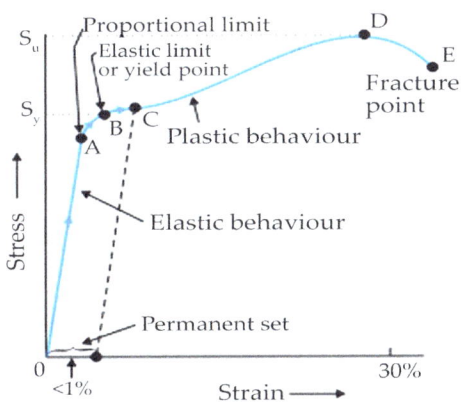

Figure 7.4. Stress-strain curve of the ductile material.

Source: https://www.toppr.com/guides/physics/mechanical-properties-of-solids/hookes-law-and-stress-strain-curve/.

7.5.1. Linear Elastic Region and the Elastic Constants

As can be viewed in the figure, the stress and strain primarily increase with the linear relationship. This is a linear elastic section of the curve and it specifies that no plastic distortion has taken place. In this particular region of the curve, with the reduced stress, the material will come back to its original shape. In this specific linear region, the line observes the association defined as *"Hooke's Law"* where the ratio of the stress to the strain is constant (Cheng et al., 2005).

The slope of the line in this specific region where the stress is proportional to the strain and is known as Young's modulus or modulus of elasticity. The modulus of elasticity outlines the properties of the material as it bears stress, deforms, and then comes back to its original shape when the stress is released. It is the measure of the stiffness of the given material. To calculate Young's modulus, divide the stress by strain in the given material. As the strain has no units, the modulus will possess the same units as that of stress, like kpi or MPa. The Young's modulus applies particularly to the situation of a component being elongated with the tensile force. This modulus is vital when it is essential to calculate how much wire or rod stretches under the tensile load (Bax and Müssig, 2008).

There are various types of moduli dependent on the method the material is being elongated, bent, or distorted. When the component suffers pure shear, for example, the cylindrical bar under some torsion, the shear modulus defines the relationship of linear-elastic stress-strain.

The axial strain is at all times escorted by lateral strains of the opposite sign in two directions jointly perpendicular to axial strain. Strains that outcome from the increase in length is labeled as positive and those that outcome in a decrease in the length is labeled as negative.

7.5.2. Poisson's Ratio

Poison's ratio is stated as negative (-) of the ratio of lateral strain to axial strain for the uniaxial stress state.

$$v = \frac{\varepsilon_{lateral}}{\varepsilon_{axial}}$$

Poisson's ratio is at times also described as the ratio of absolute values of axial and lateral strain. This ratio, just like strain, has no unit as both strains have no unit. For the stresses within the elastic range, this specific ratio is almost constant. For the perfectly isotropic elastic material, this ratio is 0.25; however, for most of the materials, the value of this ratio lies in the range of 0.28 to 0.33. Normally for the steels, Poisson's ratio will possess a value of around 0.3.

Only two of elastic constants are usually independent so if two constants are known, then using the formula below the third one can be calculated:

E = 2 (1 + n) G.

where: E = modulus of elasticity; n = Poisson's ratio; G = modulus of rigidity.

A couple of extra elastic constants that might be encountered comprise the bulk modulus (K) and the Lame's constants (m and l). The bulk modulus (K) is used to describe the state where a piece of the material is subjected to pressure increase on all sides. The relationship between the variation in pressure and the subsequent strain yielded is the bulk modulus (K). Lame's constants are generally derived from Poisson's ratio and modulus of elasticity (Andjelković et al., 2006).

7.5.3. Yield Point

In the ductile materials, the curve of stress-strain deviates at some point from the straight-line association and the Law no longer holds, since the strain increases more rapidly as compared to the stress. From this particular point on in the tensile test, some everlasting deformation takes place in the sample and the material reacts plastically to any added increase in the load or stress. The material will not come back to its original, relaxed situation when the load is released. In the brittle materials, no or little plastic deformation takes place and the material breaks near the end of a linear-elastic section of the curve.

With most of the materials, there is a slow transition from elastic to plastic behavior and the precise point at which the plastic deformation starts to take place is difficult to determine. Thus, several criteria for the beginning of yielding are utilized depending on the sensitivity of the strain measurements and the intended utilization of the data. For most of the engineering design and requirement applications, yield strength is used. The yield strength is stated as the stress needed to produce a minor, amount of the plastic deformation. The offset yield strength is stress conforming to the joining of the stress-strain curve and the line parallel to an elastic portion of the curve offset by the specified strain (Waite et al., 2009).

To decide the yield strength using this offset, a point is determined on a strain axis of 0.002 and then the line parallel to a stress-strain line is generally drawn. This line will join the stress-strain line marginally after it starts to curve and that intersection point is stated as yield strength with the 0.2% offset. A better way of observing at the yield strength with offset is that after the sample has been loaded to 0.2% offset yield strength and unloaded it will be 0.2% longer as compared to before the test. Although the yield strength represents the precise point at which the given material becomes everlastingly deformed, 0.2% elongation is thought to be an endurable amount of the sacrifice for ease it produces in describing the yield strength.

Some materials like soft copper or gray cast iron show fundamentally no linear-elastic behavior. For these kinds of materials, the typical practice is to describe the yield strength as stress needed to yield some total amount of the strain.

7.5.4. True Elastic Limit

It is considered a very low value and is linked to the motion of the few hundred dislocations. Micro strain measurements are needed to detect the strain on order of around $2 \times 10{-6}$ in/in.

7.5.5. Proportional Limit

It is the highest stress at which the stress becomes directly proportional to the strain. It is attained by perceiving the deviation from the straight-line section of the stress-strain curve.

7.5.6. Elastic Limit

It is the maximum stress that the given material can bear without any measurable everlasting strain remaining on the entire release of load. It is determined by utilizing the tiresome incremental loading-unloading test process. With the sensitivity of the strain measurements normally employed in the engineering studies, the elastic limit is far greater as compared to the proportional limit. With the escalating sensitivity of the strain measurement, the value of the elastic limit decreases till it ultimately equals the value of the true elastic limit defined from the microstrain measurements.

7.5.7. Yield Strength

It is the stress needed to create a small-particular amount of plastic deformation. The yield strength acquired by an offset approach is commonly utilized for engineering purposes as it averts the practical complications of measuring the proportional limit or elastic limit.

7.5.8. Ultimate Tensile Strength (UTS)

The UTS (ultimate tensile strength) or simply the tensile strength is the utmost engineering stress level extended in the tension test. The strength of the material is its ability to resist the external forces without fracturing. In the brittle materials, the UTS will at the end of the linear-elastic section of a stress-strain curve or near to the elastic limit. In the ductile materials, the UTS will be quite outside of the elastic portion and into the plastic portion of the stress-strain curve.

On the curve of stress-strain above, the UTS is the utmost point where a line is temporarily flat. As the UTS is centered on engineering stress, it is frequently not the same as breaking strength. In the ductile materials strain toughening takes place and stress will carry on increasing until the fracture takes place, but the curve of engineering stress-strain may exhibit decay in stress level before the fracture occurs. This is the outcome of the engineering stress being centered on the original cross-sectional area and not considering for necking that normally takes place in the test sample. The UTS may not be entirely representative of the highest level of the stress that the material can bear, but the value is not typically utilized in the design of the components anyway. For the ductile metals, the present design exercise is to use the yield strength for the sizing static components. However, as the UTS is quite easy to define and very reproducible, it is beneficial for the dedications of specifying the material and for quality control purposes (Hsiao et al., 1999).

7.6. COMPRESSIVE, BEARING, AND SHEAR PROPERTIES

7.6.1. Compressive Properties

In theory, a compression test is the opposite of the tension test concerning the direction of the loading. In this test, the specimen is squeezed whereas the displacement and the load are recorded. Compression tests outcome in the mechanical properties that comprise the compressive ultimate stress, compressive yield stress, and the compressive modulus of elasticity (LeRoux et al., 1999; Thompson et al., 2003).

Compressive yield stress is generally measured in a way alike to that done for the tensile yield strength. When testing the metals, it is stated as the stress conforming to 0.002 in./in. plastic strain. For the plastics, the compressive yield stress is usually measured at the point of the permanent produce on the curve of stress-strain. Moduli are normally greater in the compression for most of the normally used structural materials (Côté et al., 2006; Felfel et al., 2011).

Ultimate compressive strength is the stress needed to rupture the sample. This value is quite harder to conclude for the compression test as compared to conclude for the tensile test as numerous materials do not exhibit quick fracture in compression. Materials like most of the plastics that don't break can have their outcomes reported as compressive strength at the specific deformation like 1%, 5%, or 10% of the specimen's original height (Tanaka et al., 2004; Nasser and James, 2008).

For some of the materials, like concrete, compressive strength is the most vital material property that the engineers utilize when designing and building the structure. Compressive strength is commonly utilized to determine whether the concrete mixture meets the necessities of job specifications (Channell and Zukoski, 1997; Queheillalt et al., 2008).

7.6.2. Compression Test

This is any test where the material suffers opposite forces that push inwards upon the sample from the opposite sides or is compressed, squashed, crushed, or flattened. The test specimen is usually placed in between 2 plates that deal out the load applied across the complete surface area of 2 opposite faces of the test specimen and then plates are pushed together by the universal test machine triggering the sample to compress. The compressed sample is normally shortened in direction of the forces applied and also expands in a direction perpendicular to the applied force. The compression test is the opposite of a tension test (Figure 7.5).

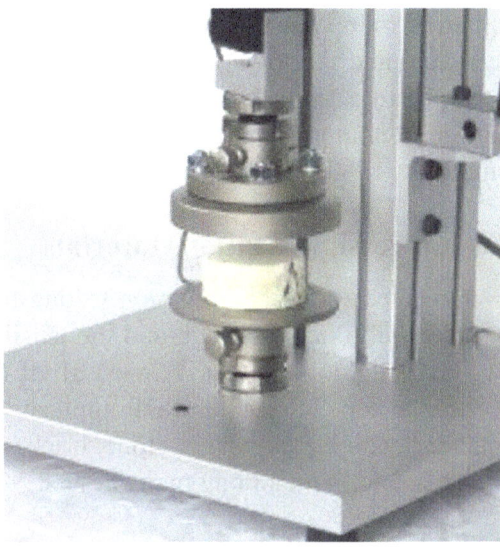

Figure 7.5. Compression test setup.

Source: https://www.testresources.net/applications/test-types/compression-test/.

7.6.3. Purpose of the Compression Test

The objective of the compression test is to determine the response or behavior of the material whereas it experiences the compressive loads by measuring the fundamental variables, like, stress, strain, and deformation. By testing the material in compression, the yield strength, compressive strength, elastic limit, ultimate strength, and the elastic modulus between other parameters might all be determined. With the understanding of these distinct parameters and values linked with the specific material, it might be concluded whether the material is or is not suited for specific applications.

7.6.4. Types of the Compression Tests

Generally, the compression test for the material involves two opposing forces directed towards one another applied to the opposite face of a test specimen so that the specimen is compressed. However, there are several different dissimilarities to this fundamental test setup that include any combination of the different variables. The common compression tests include forces applied to more axis of the sample along with the testing of the specimen at elevated and the lowered temperatures. Uniaxial, biaxial, triaxial, elevated temperature, cold temperature, creep, and fatigue are all examples of the different compression tests which may be executed upon the material (Chang et al., 1980).

7.6.5. Types of Compression Testing Materials

Usually, materials subjected to the compression testing have compressive strength normally accepted to be quite high and the tensile strength that is well thought out to be of the lower value. Almost all of the materials can suffer compressive forces depending upon their applications, but the most usual materials concrete, wood, composites, stone, brick, grouts, polymers, mortars, plastics, metals, and foam among many others.

7.6.6. Test Standards

- ASTM D575 compression test of the rubber.
- ASTM D695 compression testing for the rigid plastics.
- ASTM D6641 compression testing for the polymer matrix composite laminates.
- ASTM D7137 compressive residual strength test equipment for the damaged polymer matrix composite plates.
- ASTM E9 compression testing of the metallic materials at the room temperature.
- ASTM D905 wood adhesive bonds in shear by the compression loading.
- ISO 14126 compression fiber-reinforced plastic composites test machine.
- ISO 604 compressive plastics testing equipment.
- ISO 1856 flexible cellular polymeric materials compression EN.
- ISO 844 compressive strength of the rigid cellular plastics.

7.6.7. Common Applications

- CFD (compression force deflection) of the flexible polyurethane foam per ASTM D3574 test C.
- Compression test for the steel stress-strain | research spotlight.

- Compression test equipment and processes for the cellular foam.
- Compression testing machine – biomedical hydrogels | research spotlight.
- Compressive strength testing of the dimension stone.
- Compression testing of the hydraulic cement mortars.
- Electronic keypad and LCD compression testing.
- Gasket materials compression test equipment.
- 4 steps to choose compression test equipment for the metals.
- How to choose compression test equipment for the ceramics.
- Plastics composites compression test equipment.
- Plastic composite compression testing accessories.
- Plastics compression testing.
- Rigid cellular plastic compression properties.
- Proppant crush test for the frack sand, resin-coated sand and the ceramic proppant.
- Rubber and elastomer static and dynamic compression.
- Compression test of the electronic components.
- Rubber deflection tests in the compression per ASTM D575 & ISO 7743.
- Electronic circuit board and the components testing.

7.6.8. Bearing Properties

Bearing properties are used when mechanically designing the fastened joints. The bearing test aims to decide the deformation of the hole as the function of applied bearing stress (Erguler and Ulusay, 2009). The test sample is fundamentally a piece of plate or sheet with a cautiously prepared hole some distance from the edge. Edge to the hole diameter ratios of around 1.5 & 2.0 is common (Jakobsen et al., 2000). A hardened pin is injected through the hole and the axial load applied to a sample and the pin. The bearing stress is calculated by dividing the applied load to the pin, which ensures against the edge of the hole, by bearing area (Plaut et al., 1975; Bakar et al., 2003). Bearing produce and ultimate stresses are attained from the bearing tests. BYS is calculated from the curve of bearing stress deformation by drawing the line parallel to the initial slope usually at an offset of 0.02 times the diameter of the pin (Akizuki et al., 1986; Popescu et al., 2005).

7.6.9. Shear Properties

The shearing stress generally acts parallel to stress plane, while a tensile or the compressive stress acts normal to stress plane. Shear properties are mainly utilized in the design of the mechanically fastened components, torsion members and webs, and many other components subject to the parallel, opposing loads (Thomas and Gibbons, 1979; Sun et al., 1997). Shear properties are normally dependant on the type of shear test and there is a huge variety of distinct standard shear tests that can be executed including the blanking-shear test, single-shear test, double-shear test, torsion-shear test, and many others (MacKay and Maier, 1982; Bhowal et al., 1990).

7.7. MEASURES OF DUCTILITY

The ductility of the material is a measure of the degree to which the material will deform before the fracture. It is a significant factor when considering creating operations like rolling and extrusion (Servi and Grant, 1951; Ibrahim et al., 1972). It also offers a sign of how visible the overload damage to the component might become before fracturing. Ductility is also utilized as the quality control measure to evaluate the level of contaminations and the proper processing of the material (Nathal et al., 1982; McKamey et al., 1992) (Figure 7.6).

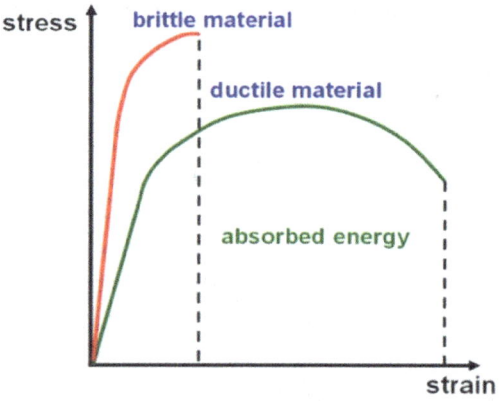

Figure 7.6. Stress-strain curves of ductile and brittle materials.

Source: https://sites.google.com/site/polymorphismmyhomepage/investigating-material-failures.

The traditional measures of the ductility are engineering strain at the fracture and reduction of the area at fracture. These properties are attained by fitting the sample back together after the fracture and measuring the variation in cross-sectional area and length. Elongation is changed in the axial length divided by the original length of the sample or portion of the specimen. It is expressed as the percentage. Because a considerable fraction of plastic deformation will normally be concentrated in a necked portion of the tensile sample, the elongation value will be dependent on gage length over which measurement is taken. The smaller the length of gage the greater will be the large localized strain in the necked region. Thus, when reporting values of the elongation, the length of gage must be given (Yun and DiCarlo, 1999).

One way to avert the difficulty from the necking is to base elongation measurement on uniform strain out to point at which the necking begins. This functions well occasionally but some curve of engineering stress-strain are frequently quite flat in the vicinity of the maximum loading and it becomes difficult to accurately develop the strain when the necking starts to happen.

The reduction of the area is changed in the cross-sectional area divided by the original cross-sectional area. This variation is measured in the necked-down section of the specimen. Like the elongation, it is normally expressed as the percentage.

As discussed previously, tension is just one way that the material can be loaded. Other methods of loading the material include bending, compression, shear, and torsion and there exist plenty of standard tests established to describe how the material acts under these conditions of other loadings.

7.8. CREEP AND THE STRESS RUPTURE PROPERTIES

7.8.1. Creep Properties

Creep is time-dependent deformation of the material whereas under the load applied that is quite below its yield strength. It is most frequently occurring at the elevated temperature but some of the materials creep at room temperature. Creep terminates in separation if steps are not taken to bring to the halt.

Creep data for design utilization are normally attained under conditions of the constant uniaxial loading and the constant temperature. Outcomes of the tests are normally plotted as the strain vs. time up to separation. As specified in an image, the creep often occurs in three stages. In the preliminary stage, strain takes place at a comparatively quick rate but the rate slowly decreases till it becomes almost constant during the second stage. This creep rate is known as a minimum creep rate or the steady-state creep rate as it is the unhurried creep rate during the entire test. In the third stage, the strain rate increases until the failure occurs (EVans and Charles, 1976) (Figure 7.7).

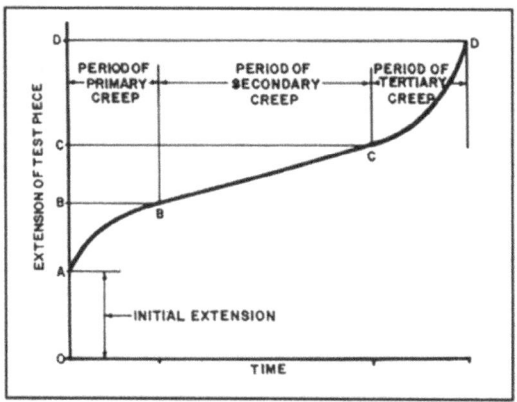

Figure 7.7. Stress-strain curve for the creep phenomenon.

Source: https://www.nationalboard.org/Index.aspx?pageID = 181.

Creep in service is generally affected by varying conditions of temperature and loading and the number of probable stress-temperature-time amalgamations is infinite. While most of the materials are subjected to creep, the mechanisms of creep are frequently different between plastics, metals, rubber, and concrete.

7.8.2. Stress Rupture Properties

Stress rupture testing is remarkably similar to the creep testing apart from that stresses are higher as compared to those utilized in the creep testing. Stress rupture tests are utilized to decide the time mandatory to yield failure so the stress rupture testing is done every time until failure. Data are plotted as log-log given in the chart above (Henriksson et al., 2008). A best-fit curve or straight line is normally attained at each temperature of the interest. This information can be utilized to infer time to the failure for longer times. A typical set of stress rupture curves is given below (Ritchie, 2011) (Figure 7.8).

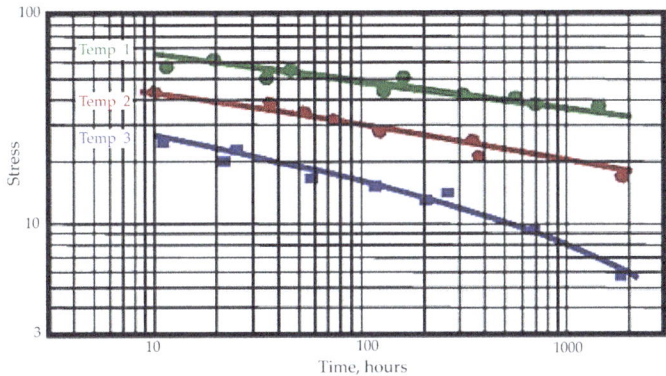

Figure 7.8. Stress ruptures properties at different temperatures.

Source: https://www.nde-ed.org/EducationResources/CommunityCollege/Materials/Mechanical/Creep.htm.

7.9. TOUGHNESS

The ability of the metal to distort plastically and to absorb the energy in the process before the fracture is termed as toughness. The focus of this definition must be placed on the ability of the metal to absorb energy before the fracture. Remember that ductility is the measure of how much anything deforms plastically before the fracture but just because the material is ductile does not make it tough. The main thing to toughness is a good combination of ductility and strength (Lawn and Marshall, 1979). The material with high ductility and high strength will possess more toughness than the material with high ductility and low strength. Thus, one method to measure the toughness of the material is by computing the area under a stress-strain curve from the tensile test. This value is known as material toughness and has the units of energy per volume. The toughness of material equates to slow absorption of the energy by the material (Lowhaphandu and Lewandowski, 1998) (Figure 7.9).

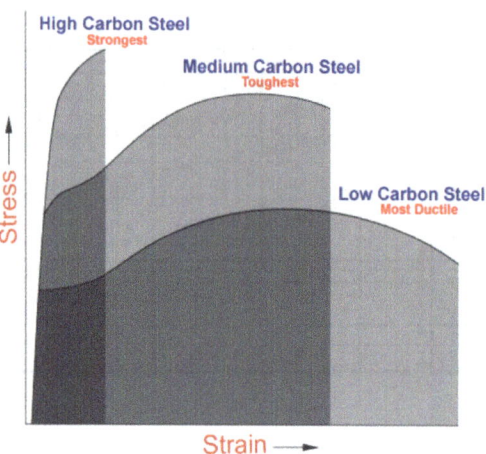

Figure 7.9. The toughness of the different materials.

Source: https://www.nde-ed.org/EducationResources/CommunityCollege/Materials/Mechanical/Toughness.htm.

There are various variables that have a profound impact on the toughness of the material. These variables are as:

- Strain rate;
- Temperature;
- Notch effect.

A metal might possess suitable toughness under the static loads but might fail under dynamic impact or loads. As a principle ductility and thus toughness decrease as the rate of the loading increases. Temperature is the variable to have a major impact on its toughness. As the temperature is lowered, toughness, and ductility also decrease. The third variable is called the notch effect which deals with the distribution of the stress. A material might exhibit good toughness when the stress applied is uniaxial but when the multiaxial stress state is formed because of the existence of notch, the material may not bear the simultaneous plastic and elastic deformation in the several directions (Wong et al., 1997).

There are various standard kinds of toughness tests that create data for particular loading conditions or component design methods. The three toughness properties discussed in detail are as:

- Impact toughness;
- Notch toughness; and
- Fracture toughness.

7.9.1. Impact Toughness

The impact toughness of the material can be decided with an Izod or Charpy test. These tests are termed after their originators and were established in the early 1900s before the theory of fracture mechanics was available. Impact properties aren't directly utilized in the fracture mechanics calculations but the economic influence tests continue to be applied as the quality control approach to assessing the notch sensitivity and for associating the comparative toughness of the engineering materials (Tan et al., 2005; Lin et al., 2008) (Figure 7.10).

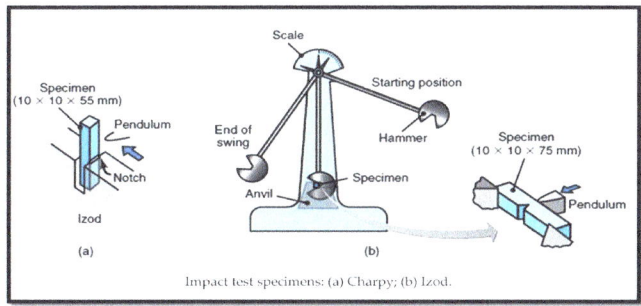

Impact test specimens: (a) Charpy; (b) Izod.

Figure 7.10. Izod and Charpy impact test setup.

Source: https://www.quora.com/How-do-Charpy-and-Izod-impact-tests-differ.

The two tests utilize different samples and approaches to holding the samples, but both these tests utilize the pendulum-testing machine. For both of these tests, the sample is broken by the single overload event because of the impact of a pendulum. The stop pointer is utilized to record how far the pendulum swings back up just after fracturing the sample. The impact toughness of the metal is concluded by measuring the absorbed energy in fracture of the sample (Nagendra et al., 2000; Ma et al., 2005). This is simply attained by noticing the height at which this pendulum is unconfined and height to which this pendulum swings just after it has hit the specimen. The height of pendulum times the weight of pendulum yields the potential energy and difference in the potential energy of pendulum at the start and the end of the test is just equal to the energy absorbed (Bhadeshia et al., 1995; Yan et al., 2006) (Figure 7.11).

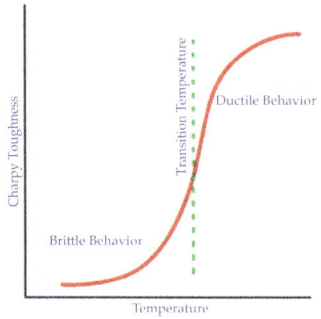

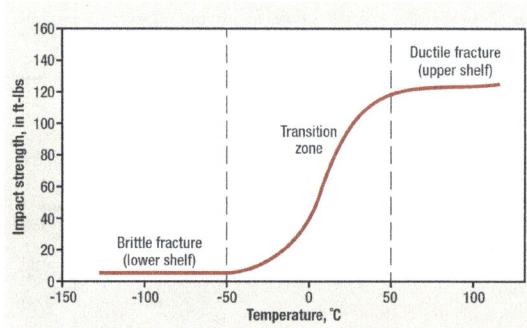

Figure 7.11. Influence of temperature on the impact toughness.

Source: https://www.industrialheating.com/articles/92389-impact-testing-part-1.

Toughness is significantly affected by the temperature, an Izod or Charpy test is frequently repeated several times with each sample tested at a different temperature. This yields a graph of the impact toughness for material as the function of temperature. An impact toughness vs. temperature graph for the steel is displayed in the image (Stolyarov et al., 2006; Liao et al., 2009). It can be viewed that at the low temperatures the given material is more brittle, and the impact toughness is considerably low. At the high temperatures, the given material is more ductile and the impact toughness is quite higher. The transition temperature is a boundary between ductile and brittle behavior and the transition temperature is frequently an extremely significant consideration in the choice of the material (Svensson and Gretoft, 1990; Song et al., 2010).

7.9.2. Notch-Toughness

This is the ability that the material must absorb the energy in the existence of a flaw. As previously mentioned, in the existence of a flaw, like a crack or not, a material will possibly display a lower level of toughness. When the flaw is existent in the material, loading encourages the triaxial tension stress state near to the flaw (Lowhaphandu and Lewandowski, 1998; Lehmann et al., 2003). The material advances the plastic strains as yield stress is surpassed in the region adjacent to the crack tip. However, the amount of plastic deformation is limited by the surrounding material, which stays elastic. When the material is averted from deforming plastically, the material fails in a brittle manner (Norris Jr, 1979; Wesseling et al., 2004).

Notch-toughness is generally measured with a variety of samples like the Charpy V-notch impact sample or dynamic tear test sample. As with the regular impact testing, these tests are frequently repeated several times with samples tested at a different temperature. With these samples and by changing the temperature and loading speed, it is probable to produce curves like those displayed in the graph. Typically, only impact and static testing are directed but it must be recognized that numerous components in the service realize intermediate loading rates in the range of the red line (Leis et al., 1998; Baczynski et al., 1999) (Figure 7.12).

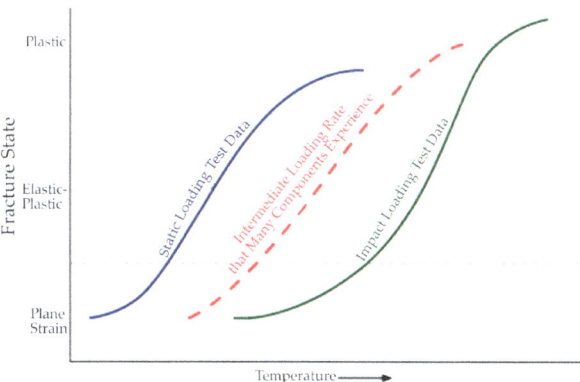

Figure 7.12. Fracture states in the notch impact test concerning the temperature variation.

Source: https://www.nde-ed.org/EducationResources/CommunityCollege/Materials/Mechanical/NotchToughness.htm.

7.9.3. Fracture Toughness

This is a sign of the amount of the stress needed to propagate the preexisting flaw. It is very significant material property as the existence of flaws isn't entirely avoidable in the fabrication, processing, or service of the material. Flaws may look like cracks, metallurgical inclusions, voids, design discontinuities, weld defects, or some amalgamation thereof (Claussen, 1976; Low and Brown, 1981). As engineers can never be completely sure that the material is flaw-free, it a common practice to suppose that the flaw of some selected size will be existent in a few numbers of the components and utilize the LEFM (linear elastic fracture mechanics) method to design the critical components. This approach utilizes the features and flaw size, loading conditions, component geometry and the material property known as fracture toughness to assess the ability of the composition comprising a flaw to the resist fracture (EVans and Charles, 1976) (Figure 7.13).

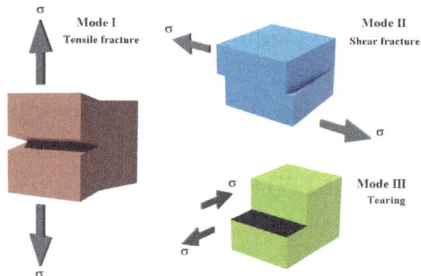

Figure 7.13. Different kinds of fracture modes.

Source: https://www.researchgate.net/figure/Fracture-modes-The-fracture-toughness-K-IC-is-a-threshold-property-so-that_fig6_221914692.

A parameter known as the stress-intensity factor (K) is utilized to conclude the fracture toughness of several materials (Gojny et al., 2004). The Roman numeral subscript specifies a mode of the fracture and three modes of the fracture are demonstrated in the image to the right. The mode I fracture is a situation in which the crack plane is usually normal to the direction of the largest tensile loading. This is the commonly faced mode and, thus, for the remnants of material K_I is considered (Anstis et al., 1981; Zhang et al., 2014). The stress intensity factor is the function of crack size, loading, and structural geometry. The factor of stress intensity may be signified by the equation below (Ritchie et al., 1973; Hahn and Rosenfield, 1975):

$$K_I = \sigma\sqrt{\pi a \beta}$$

Where: $K_I \rightarrow$ fracture toughness in $MPa\sqrt{m}\,(psi\sqrt{in})$

s $\rightarrow$ applied stress in Psi or MPa;

a $\rightarrow$ crack length in inches or meters;

B $\rightarrow$ component geometry factor and crack length that is different for every sample and is dimensionless.

7.9.3.1. Role of the Material Thickness

Samples having standard ratios, but diverse absolute sizes yield different values for the K_I. This outcome since the stress states near to flaw changes with the sample thickness (B) till the thickness surpasses some critical dimension. Once this thickness surpasses the precarious dimension, the value of the K_I becomes comparatively constant and the value, K_{IC}, is the true material property which generally is known as plane-strain fracture toughness (Jeffery, 1921; Rice and Rosengren, 1968). The relationship amongst stress intensity K_I and the fracture toughness K_{IC} is comparable to the association between tensile stress and stress. The stress intensity K_I signifies the level of stress at the tip of the crack and fracture toughness K_{IC} is the highest value of the stress intensity that the material under very particular conditions that the material can bear without fracture. As the factor of stress intensity reaches the K_{IC} value, unstable fracture takes place. As with the other mechanical properties of a material, K_{IC} is usually reported in the reference books and some other sources (Runesson et al., 1991; Eraslan and Akis, 2006) (Figure 7.14).

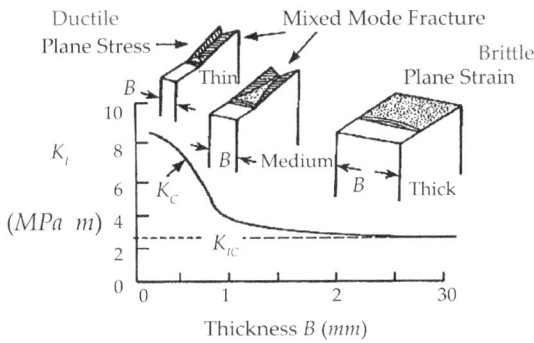

Figure 7.14. Effect of sample thickness on the fracture toughness.

7.9.3.2. Plane Strain and Plane Stress

When the material having a crack is loaded in the tension, the materials produce plastic strains as yield stress is surpassed in the region adjacent to the crack tip. Material within the crack tip stress field, located close to the free surface, can laterally deform as there will be no stresses normal to a free surface. The state of the stress inclines to biaxial and material fractures in the characteristic ductile manner, with 45° shear lip being developed at every free surface (McClung et al., 1991; Anthoine, 1997). This condition is known as plane-stress and it takes place in comparatively thin bodies where stress through the thickness cannot vary appreciably because of the thin section (Figure 7.15).

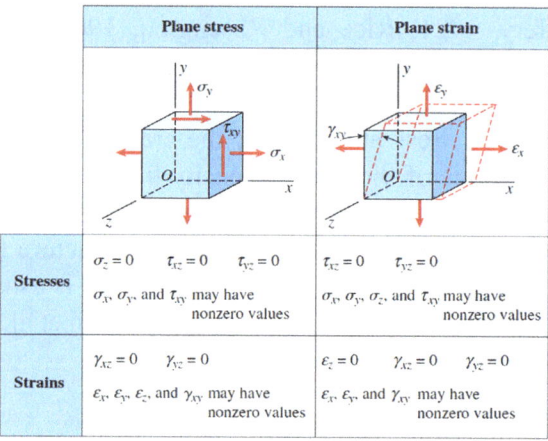

Figure 7.15. Characteristics of the plane strain and plane stress condition.

However, materials with no free surfaces of close thick component are not free to laterally deform as it is reserved by the neighboring material. The state of stress under these conditions inclines to the triaxial and there is 0 strain perpendicular to both stress axis and direction of the crack propagation when the material is loaded in the tension. This condition is known as plane strain and is discovered in thick plates (Murrell, 1964).

7.9.3.3. Plane Strain Testing of Fracture Toughness

When performing the fracture toughness test, the common test sample configurations are the SENB (single edge notch bend), and the CT (compact tension) specimens. From the discussion above, it is quite clear that the precise determination of plane-strain fracture toughness needs the sample whose thickness surpasses some critical thickness (B) (Figure 7.16). Testing has displayed that plane-strain conditions normally prevail when:

$$B \geq 2.5 \left(\frac{K_{IC}}{\sigma_y} \right)^2$$

where:

B → the minimum thickness that develops the condition where the energy of plastic strain at the tip of crack is minimal;

K_{IC} → fracture toughness of the material;

s_y → yield stress of the material.

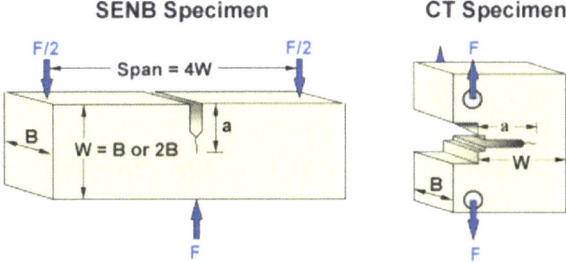

Figure 7.16. Specimen specifications for the plane strain fracture toughness.

Source: https://www.totalmateria.com/page.aspx?ID=CheckArticle&site=kts &NM=294.

When a material of unidentified fracture toughness is tested, the sample of full material segment thickness is analyzed or the sample is sized centered on the prediction of fracture toughness. If the value of fracture toughness resulting from the test does not satisfy the condition of the equation above, the test should be repeated using the thicker specimen. In addition to the thickness calculation, the test specifications have various other requirements that should be met before the test can be considered to have an outcome in the K_{IC} value.

When the test fails to encounter the thickness and some other test conditions that are in the place to assure plane-strain condition, the values of fracture toughness developed are given the title K_C. Sometimes it is not probable to produce the sample that meets the requirement of thickness. For instance, when a comparatively thin plate product with the high toughness is being tested, it may not be feasible to yield a thicker specimen with the plain-strain conditions at the tip of the crack (Fernando and Williams, 1980).

7.9.3.4. Plane Stress and the Transitional Stress States

In circumstances where plastic energy at the tip of the crack is not negligible, the parameters of other fracture mechanics, like the R-curve or J integral, can be employed to characterize the material. The toughness data yielded by the other tests will usually be dependent on the thickness of the tested product and will not be the true material property. However, conditions of plane-strain do not exist in all the structural configurations and utilizing K_{IC} values in the design of the comparatively thin areas might outcome in excess conservatism and the cost or weight penalty. In situations where the state of actual stress is the plane-stress or more generally some intermediate or the transitional stress state, it is suitable to use R-curve or J integral data, which account for the slow, stable fracture rather than the rapid fracture.

7.9.3.5. Uses of the Plane Strain Fracture Toughness

K_{IC} values are utilized to conclude the length of critical crack when the given stress is normally applied to the component.

$$\sigma_C \leq \frac{K_{IC}}{Y\sqrt{\pi a}}$$

where:

$s_c \rightarrow$ precarious applied stress that will normally cause failure.

$K_{IC} \rightarrow$ plane strain fracture toughness.

$Y \rightarrow$ constant associated with the geometry of the sample.

$a \rightarrow$ crack length for the edge cracks or the length of one-half crack for the internal crack.

K_{IC} values are also utilized to compute the value of critical stress when the crack of the given length is discovered in the component.

$$a_C = \frac{1}{\pi}\left(\frac{K_{IC}}{\sigma Y}\right)^2$$

where:

a → Crack length for the edge cracks or the length of one-half crack for the internal crack.

s → Applied stress to the material.

K_{IC} → Plane strain fracture toughness.

Y → Constantly linked to the geometry of the sample.

7.9.3.6. Grain Direction and the Fracture Orientation

The fracture toughness of the material commonly changes with the grain direction. Thus, it is usual to specify sample and crack directions by the ordered pair of grain direction symbols. The first letter describes the grain direction normal to crack plane. The second letter specifies the grain direction parallel to the fracture plane. For the flat sections of several products, e.g., plate, forgings, extrusions, etc., in which 3 grain directions are specified (T) transverse, (L) longitudinal, and (S) short transverse, the 6 chief fracture path directions areas: L-T, T-L, L-S, T-S, S-L & S-T (Figure 7.17).

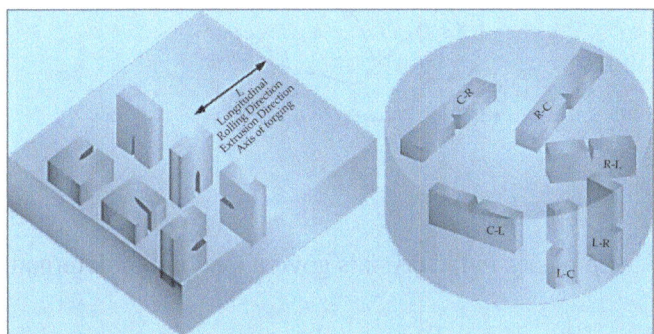

Figure 7.17. Rolling direction versus fracture orientation.

Source: https://www.nde-ed.org/EducationResources/CommunityCollege/Materials/Mechanical/FractureToughness.htm.

7.10. PLASTIC/ELASTIC DEFORMATION

When an adequate load is applied to a metal or some other structural material, the load will change the shape of the material. The change in shape is known as deformation. A momentary change in shape that is self-reversing when the force is released, so that the object comes back to the original shape, is known as elastic deformation. In other words, the elastic deformation is the change in the shape of the material at quite low stress that can be recovered after the stress is released. This kind of deformation comprises elongating of the bonds, but the atoms don't slip past one another (Figure 7.18).

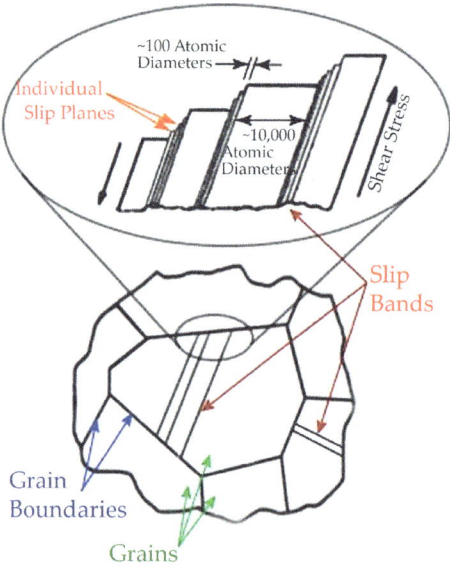

Figure 7.18. Slip planes in the crystals govern the kind of deformation (plastic or elastic).

Source: https://www.nde-ed.org/EducationResources/CommunityCollege/Materials/Structure/deformation.htm.

When the stress is adequate to perpetually distort the metal, it is known as plastic deformation. As discussed in the section on the crystal defects, the plastic deformation comprises the breaking of the limited number of the atomic bonds by the movement of the dislocations. Remember that the force required to fracture the bonds of all atoms in the crystal plane suddenly is very large. Though, the movement of dislocations allows atoms in the crystal planes to slip past each other at the much lower levels of stress. As the energy needed to move is the lowest along densest planes of the atoms, dislocations have the preferred direction of the travel within the grain of the material. This outcome in a slip that takes place along the parallel planes within the grain. The parallel slip planes combine to create slip bands, which generally can be viewed with an optical microscope. The slip band looks like the single line under a microscope but it is made up of nearly spaced parallel slip planes as exhibited in the image.

7.11. FATIGUE

In the materials science, the fatigue is weakening of the material caused by the cyclic loading that outcomes in localized and progressive structural damage and growth of the cracks (Eskandari and Kim, 2017; Burhan and Kim, 2018). Once the crack has started, each cycle of loading will advance the crack a small amount, normally producing corrugations on some parts of the fracture surface (Stephens et al., 2000; Sosnovskiy, 2003). The crack will grow till it reaches the critical size, which takes place when the factor of stress intensity of the crack surpasses the fracture toughness of the material, producing quick propagation and normally complete fracture of structure (Bathias, 1999; Shigley et al., 2007).

Fatigue has conventionally been linked with failure of the metal components which outlines the term of metal fatigue. In the 19th century, sudden failure of the metal railway axles was considered to be caused by metal *crystallizing* due to the brittle look of fracture surface but this has been disproved. Most materials appear to suffer some kind of fatigue associated with failures like plastics, composites, and ceramics (Vladimir and Imanaka, 1997; Kukharev, 2015).

If loads are above the certain threshold, the microscopic cracks will start to *initiate* at the stress concentrations like holes, PSBs (*persistent slip bands*), composite interfaces or the grain boundaries in the metals The nominal values of maximum stress that cause the damage might be quite less as compared to the strength of material, usually quoted as UTS (ultimate tensile stress) limit or yield stress limit (Elberl, 1971; Brown and Miller, 1973).

7.11.1. Stages of Fatigue

Historically, fatigue has normally been separated into the regions of the *high cycle fatigue* that need more than 10000 cycles to failure where the stress is low and mainly elastic and the low cycle fatigue where there is noteworthy plasticity. Experiments have exhibited that low cycle fatigue is a crack growth (Schijve, 1978, 2003; Freund and Suresh, 2004).

Fatigue failures, both for low and high cycle, all of this follow the same fundamental steps procedure of crack commencement, stage I growth of crack, stage II growth of crack, and finally eventual failure. To begin the procedure cracks should nucleate within the material. This process can take place either at the stress risers in the metallic samples or at the areas with high void density in the polymer samples. These cracks propagate gradually at first during the *stage I* growth of crack along the crystallographic planes, where the shear stresses are highest. Once the cracks come at a precarious size, they propagate rapidly during *stage II* growth of the crack in the direction perpendicular to an applied force.

7.11.2. Crack Initiation

The creation of preliminary cracks preceding failure of fatigue is the separate procedure comprising of four discrete steps in the metallic samples. The material will produce cell structures and then harden in reaction to the load applied. This increases the amplitude of applied stress given the new limitations on the strain. These newly created cell structures will ultimately break down with the creation of PSBs (persistent slip bands). Slip in a material is localized at persistent slip bands, and the excessive slip can now aid as the stress concentrator for the crack to form. Nucleation of the crack to noticeable size accounts for most of the cracking procedure (Weibull, 1952, 2013). PSB induced slip planes outcome in extrusions and intrusions along the surface of a material, frequently occurring in the pairs. This slip is not the microstructural change within material but instead a propagation of the dislocations within a material. Instead of the smooth interface, the extrusions and intrusions will root the surface of a material to look like the edge of the deck of cards, where all the cards are not perfectly aligned. Slip-induced extrusions and intrusions create a fine surface structures on the material. With the surface structure size inversely associated with the stress concentration factors, the PSB induced surface slip rapidly becomes the perfect place for the fractures to initiate (Forsyth, 1953; Fleck et al., 1985; Schütz, 1996).

7.11.3. Crack Growth

Most of the life of fatigue is normally spent in a crack growth phase. The growth rate is chiefly driven by the range of the cyclic loading even though extra factors like environment, mean stress, underloads, and overloads can also influence the growth rate. Crack growth might stop below a certain threshold. Fatigue cracks develop from material or developed defects. In the situation of aluminum, the cracks normally grow from the surface, where the water vapors from the atmosphere can arrive at the tip of crack and separate into the atomic hydrogen which triggers hydrogen embrittlement. Cracks progressing internally are insulated from atmosphere and advance in the vacuum where the growth rate is usually an order of the magnitude slower as compared to the surface crack (Rankine, 1843; Kim and Laird, 1978; Murakami and Miller, 2005).

When the intensity of stress exceeds the critical value called the fracture toughness, and unmaintainable fracture will take place, normally by the procedure of micro-void coalescence. Before the final fracture, the surface of fracture may comprise a mixture of fast fracture and fatigue.

7.11.4. Characteristics of the Fatigue

Fracture of the aluminum crank arm. A dark area of the striations: slow growth of the crack. Bright granular area: rapid fracture (Braithwaite, 1854; Basquin, 1910).

- In the metal alloys and for simplifying circumstances when there are no microscopic or macroscopic discontinuities, the procedure begins with dislocation movements at a microscopic level.
- Microscopic and macroscopic discontinuities along with the component design characteristics which cause stress concentrations are common positions at which the fatigue procedure begins.
- Fatigue is the procedure that has the degree of randomness, frequently exhibiting substantial scatter even in the seemingly alike samples in well-organized environments.
- Fatigue is generally linked with the tensile stresses, but cracks of fatigue have been reported because of the compressive loads.
- The greater the stress applied range, the shorter is the life.
- Fatigue life scatter inclines to increase for longer fatigue lives.
- Damage is irreversible. Materials do not recover when restored.

- Fatigue life is affected by a variety of factors, like temperature, metallurgical microstructure, surface finish, the occurrence of inert or oxidizing chemicals, scuffing contact, residual stresses, etc.

- Some materials show the theoretical fatigue limit quite below which the continued loading does not lead to fatigue failure.

- High cycle strength fatigue can be defined by the stress-based parameters. The load organized servo-hydraulic test rig is usually utilized in these tests, with the frequencies of around 20 to 50 Hz. Other sorts of machines like the resonant magnetic machines can also be utilized, to accomplish frequencies around 250 Hz.

- Low-cycle fatigue is linked with the localized plastic behavior in metals; therefore, the strain-based parameter must be utilized for fatigue life estimation in metals. Testing is accompanied by the constant strain amplitudes usually at 0.01–5 Hz.

7.11.5. Predicting Fatigue Life

The ASTM defines *fatigue life* N_f as several cycles of the stress of the specified character that the sample withstands before the failure of the specified nature occurs. For some of the materials, notably titanium and steel, there is the theoretical value for the stress amplitude below which material won't fail for any number of the cycles, known as fatigue limit, endurance limit or fatigue strength (Cadwell et al., 1940) (Figure 7.19).

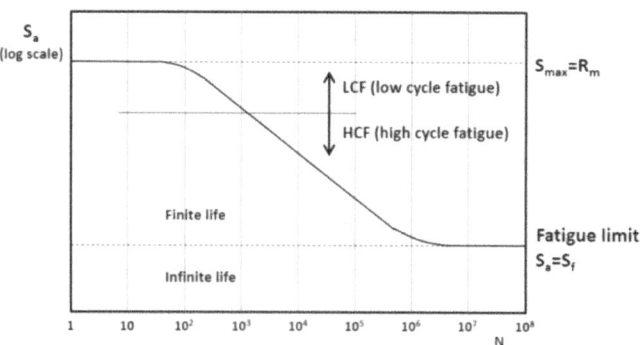

Figure 7.19. Forecasting the life of material utilizing fatigue testing.

Source: https://www.fatec-engineering.com/2018/02/20/description-of-a-s-n-curve/.

Engineers have used several approaches to conclude the fatigue life of material:

- Stress-life method;
- Strain-life method;
- Crack growth method;
- Probabilistic methods.

Whether utilizing stress/strain-life method or using the growth of crack approach, complex or the variable amplitude loading is decreased to the series of fatigue corresponding simple cyclic loadings utilizing the technique like the rain flow counting algorithm (Palmgren, 1924; Azari et al., 1984).

REFERENCES

1. Akizuki, S., Mow, V. C., Müller, F., Pita, J. C., Howell, D. S., & Manicourt, D. H., (1986). Tensile properties of human knee joint cartilage: I. Influence of ionic conditions, weight bearing, and fibrillation on the tensile modulus. *Journal of Orthopaedic Research, 4*(4), 379–392.

2. Altman, G., Horan, R., Martin, I., Farhadi, J., Stark, P., Volloch, V., & Kaplan, D. L., (2002). Cell differentiation by mechanical stress. *The FASEB Journal, 16*(2), 270–272.

3. Andjelković, M., Van Camp, J., De Meulenaer, B., Depaemelaere, G., Socaciu, C., Verloo, M., & Verhe, R., (2006). Iron-chelation properties of phenolic acids bearing catechol and galloyl groups. *Food Chemistry, 98*(1), 23–31.

4. Anstis, G. R., Chantikul, P., Lawn, B. R., & Marshall, D. B., (1981). A critical evaluation of indentation techniques for measuring fracture toughness: I, direct crack measurements. *Journal of the American Ceramic Society, 64*(9), 533–538.

5. Anthoine, A., (1997). Homogenization of periodic masonry: Plane stress, generalized plane strain or 3D modeling? *Communications in Numerical Methods in Engineering, 13*(5), 319–326.

6. Armstrong, R. W., (1970). The influence of polycrystal grain size on several mechanical properties of materials. *Metallurgical and Materials Transactions B, 1*(5), 1169–1176.

7. Ascenzi, A., & Bonucci, E., (1967). The tensile properties of single osteons. *The Anatomical Record, 158*(4), 375–386.

8. Azari, Z., Lebienvenu, M., & Pluvinage, G., (1984). Functions of damage in low-cycle fatigue. In: *Fracture 84* (Vol. 1, pp. 1815–1821). Pergamon.

9. Baczynski, G. J., Jonas, J. J., & Collins, L. E., (1999). The influence of rolling practice on notch toughness and texture development in high-strength line pipe. *Metallurgical and Materials Transactions A, 30*(12), 3045–3054.

10. Bakar, M. A., Cheng, M. H. W., Tang, S. M., Yu, S. C., Liao, K., Tan, C. T., & Cheang, P., (2003). Tensile properties, tension-tension fatigue and biological response of polyetheretherketone–hydroxyapatite composites for load-bearing orthopedic implants. *Biomaterials, 24*(13), 2245–2250.

11. Barker, L. M., & Hollenbach, R. E., (1965). Interferometer technique for measuring the dynamic mechanical properties of materials. *Review of Scientific Instruments, 36*(11), 1617–1620.

12. Barlow, D. H., Becker, R., Leitenberg, H., & Agras, W. S., (1970). A mechanical strain gauge for recording penile circumference change. *Journal of Applied Behavior Analysis, 3*(1), 73.

13. Barthelat, F., Li, C. M., Comi, C., & Espinosa, H. D., (2006). Mechanical properties of nacre constituents and their impact on mechanical performance. *Journal of Materials Research, 21*(8), 1977–1986.

14. Basquin, O. H., (1910). The exponential law of endurance tests. In: *Proc Am Soc Test Mater* (Vol. 10, pp. 625–630).

15. Bassett, C. A. L., & Becker, R. O., (1962). Generation of electric potentials by bone in response to mechanical stress. *Science, 137*(3535), 1063–1064.

16. Bathias, C., (1999). There is no infinite fatigue life in metallic materials. *Fatigue and Fracture of Engineering Materials and Structures, 22*(7), 559–565.

17. Bax, B., & Müssig, J., (2008). Impact and tensile properties of PLA/Cordenka and PLA/flax composites. *Composites Science and Technology, 68*(7/8), 1601–1607.

18. Berthelot, J. M., & Ling, F. F., (1999). *Composite Materials: Mechanical Behavior and Structural Analysis* (Vol. 1, pp. 3–14). New York: Springer.

19. Bhadeshia, H. K. D. H., MacKay, D. J. C., & Svensson, L. E., (1995). Impact toughness of C–Mn steel arc welds – Bayesian neural network analysis. *Materials Science and Technology, 11*(10), 1046–1051.

20. Bhowal, P. R., Wright, E. F., & Raymond, E. L., (1990). Effects of cooling rate and γ′ morphology on creep and stress-rupture properties of a powder metallurgy superalloy. *Metallurgical Transactions A, 21*(6), 1709–1717.

21. Bigliani, L. U., Pollock, R. G., Soslowsky, L. J., Flatow, E. L., Pawluk, R. J., & Mow, V. C., (1992). Tensile properties of the inferior glenohumeral ligament. *Journal of Orthopaedic Research, 10*(2), 187–197.

22. Black, J., (2005). *Biological Performance of Materials: Fundamentals of Biocompatibility* (Vol. 1, pp. 1–22). Crc Press.

23. Braithwaite, F., (1854). On the fatigue and consequent fracture

of metals. In: *Minutes of the Proceedings of the Institution of Civil Engineers* (Vol. 13, No. 1854, pp. 463–467). Thomas Telford-ICE Virtual Library.

24. Braithwaite, F., (1854). On the fatigue and consequent fracture of metals. In: *Minutes of the Proceedings of the Institution of Civil Engineers* (Vol. 13, No. 1854, pp. 463–467). Thomas Telford-ICE Virtual Library.

25. Brown, M. W., & Miller, K. J., (1973). A theory for fatigue failure under multiaxial stress-strain conditions. *Proceedings of the Institution of Mechanical Engineers, 187*(1), 745–755.

26. Burhan, I., & Kim, H. S., (2018). SN curve models for composite materials characterization: An evaluative review. *Journal of Composites Science, 2*(3), 38.

27. Cadwell, S. M., Merrill, R. A., Sloman, C. M., & Yost, F. L., (1940). Dynamic fatigue life of rubber. *Industrial and Engineering Chemistry Analytical Edition, 12*(1), 19–23.

28. Carreker Jr, R. P., & Hibbard Jr, W. R., (1953). Tensile deformation of high-purity copper as a function of temperature, strain rate, and grain size. *Acta Metallurgica, 1*(6), 654–663.

29. Chang, C. D., Waki, M., Ahmad, M., Meienhofer, J., Lundell, E. O., & Haug, J. D., (1980). Preparation and properties of nα-9-fluorenylmethyloxycarbonylamino acids bearing tert.-butyl side chain protection. *International Journal of Peptide and Protein Research, 15*(1), 59–66.

30. Channell, G. M., & Zukoski, C. F., (1997). Shear and compressive rheology of aggregated alumina suspensions. *AIChE Journal, 43*(7), 1700–1708.

31. Cheng, Q., Bao, J., Park, J., Liang, Z., Zhang, C., & Wang, B., (2009). High mechanical performance composite conductor: Multi-walled carbon nanotube sheet/bismaleimide nanocomposites. *Advanced Functional Materials, 19*(20), 3219–3225.

32. Cheng, Q., Jiang, L., & Tang, Z., (2014). Bioinspired layered materials with superior mechanical performance. *Accounts of Chemical Research, 47*(4), 1256–1266.

33. Cheng, S., Ma, E., Wang, Y. M., Kecskes, L. J., Youssef, K. M., Koch, C. C., & Han, K., (2005). Tensile properties of in situ consolidated nanocrystalline Cu. *Acta Materialia, 53*(5), 1521–1533.

34. Chiquet, M., (1999). Regulation of extracellular matrix gene expression by mechanical stress. *Matrix Biology*, *18*(5), 417–426.

35. Claussen, N., (1976). Fracture toughness of Al2O3 with an unstabilized ZrO2Dispersed phase. *Journal of the American Ceramic Society*, *59*(1-2), 49–51.

36. Côté, F., Deshpande, V. S., Fleck, N. A., & Evans, A. G., (2006). The compressive and shear responses of corrugated and diamond lattice materials. *International Journal of Solids and Structures*, *43*(20), 6220–6242.

37. Craig, R. R., & Kurdila, A. J., (2006). *Fundamentals of Structural Dynamics* (Vol. 1, pp. 1–19). John Wiley & Sons.

38. Curtis, A. S. G., & Clark, P., (1990). The effects of topographic and mechanical properties of materials on cell behavior. *Crit. Rev. Bio. Compat.*, *5*, 343–362.

39. Dalla Torre, F., Van Swygenhoven, H., & Victoria, M., (2002). Nanocrystalline electrodeposited Ni: Microstructure and tensile properties. *Acta Materialia*, *50*(15), 3957–3970.

40. Davies, P. F., & Tripathi, S. C., (1993). Mechanical stress mechanisms and the cell. An endothelial paradigm. *Circulation Research*, *72*(2), 239–245.

41. Davim, J. P., (2008). *Machining: Fundamentals and Recent Advances* (Vol. 1, pp. 1–23). Springer Science & Business Media.

42. De Wolf, I., (1996). Micro-Raman spectroscopy to study local mechanical stress in silicon integrated circuits. *Semiconductor Science and Technology*, *11*(2), 139.

43. Donald, I. B., & Chen, Z., (1997). Slope stability analysis by the upper bound approach: Fundamentals and methods. *Canadian Geotechnical Journal*, *34*(6), 853–862.

44. Dowling, N. E., (2012). *Mechanical Behavior of Materials: Engineering Methods for Deformation, Fracture, and Fatigue* (Vol. 1, pp. 1–18). Pearson.

45. Drury, J. L., Dennis, R. G., & Mooney, D. J., (2004). The tensile properties of alginate hydrogels. *Biomaterials*, *25*(16), 3187–3199.

46. Duncan, J. M., & Chang, C. Y., (1970). Nonlinear analysis of stress and strain in soils. *Journal of Soil Mechanics and Foundations Div.*, *1*(1), 1–20.

47. Duncan, R. L., & Turner, C. H., (1995). Mechanotransduction and the functional response of bone to mechanical strain. *Calcified Tissue International, 57*(5), 344–358.

48. Ehrlich, P. J., & Lanyon, L. E., (2002). Mechanical strain and bone cell function: A review. *Osteoporosis International, 13*(9), 688–700.

49. Elberl, W., (1971). The significance of fatigue crack closure. *Damage Tolerance in Aircraft Structures, 486,* 230.

50. Elvin, N. G., Elvin, A. A., & Spector, M., (2001). A self-powered mechanical strain energy sensor. *Smart Materials and Structures, 10*(2), 293.

51. Eraslan, A. N., & Akis, T., (2006). On the plane strain and plane stress solutions of functionally graded rotating solid shaft and solid disk problems. *Acta Mechanica., 181*(1/2), 43–63.

52. Erguler, Z. A., & Ulusay, R., (2009). Water-induced variations in mechanical properties of clay-bearing rocks. *International Journal of Rock Mechanics and Mining Sciences, 46*(2), 355–370.

53. Eskandari, H., & Kim, H., (2017). A Theory for mathematical framework and fatigue damage function for the SN plane. In: *Fatigue and Fracture Test Planning, Test Data Acquisitions and Analysis Engineering* (Vol. 1, pp. 4–16). ASTM International.

54. EVans, A. G., & Charles, E. A., (1976). Fracture toughness determinations by indentation. *Journal of the American Ceramic Society, 59*(7/8), 371–372.

55. Faridmehr, I., Osman, M. H., Adnan, A. B., Nejad, A. F., Hodjati, R., & Azimi, M., (2014). Correlation between engineering stress-strain and true stress-strain curve. *American Journal of Civil Engineering and Architecture, 2*(1), 53–59.

56. Felfel, R. M., Ahmed, I., Parsons, A. J., Walker, G. S., & Rudd, C. D., (2011). In vitro degradation, flexural, compressive and shear properties of fully bioresorbable composite rods. *Journal of the Mechanical Behavior of Biomedical Materials, 4*(7), 1462–1472.

57. Fernando, P. L., & Williams, J. G., (1980). Plane stress and plane strain fractures in polypropylene. *Polymer Engineering and Science, 20*(3), 215–220.

58. Fink, L. H., & Carlsen, K., (1978). Operating under stress and strain. *IEEE Spectrum; (United States), 15*(3), 1–20.

59. Fleck, N. A., Shin, C. S., & Smith, R. A., (1985). Fatigue crack growth under compressive loading. *Engineering Fracture Mechanics*, *21*(1), 173–185.

60. Forsyth, P. J. E., (1953). Exudation of material from slip bands at the surface of fatigued crystals of an aluminum–copper alloy. *Nature*, *171*(4343), 172–173.

61. Freund, L. B., & Suresh, S., (2004). *Thin Film Materials: Stress, Defect Formation, and Surface Evolution* (Vol. 11, pp. 1–19). Cambridge University Press.

62. Fundenberger, J. J., Philippe, M. J., Wagner, F., & Esling, C., (1997). Modeling and prediction of mechanical properties for materials with hexagonal symmetry (zinc, titanium and zirconium alloys). *Acta Materialia*, *45*(10), 4041–4055.

63. Gojny, F. H., Wichmann, M. H. G., Köpke, U., Fiedler, B., & Schulte, K., (2004). Carbon nanotube-reinforced epoxy-composites: Enhanced stiffness and fracture toughness at low nanotube content. *Composites Science and Technology*, *64*(15), 2363–2371.

64. Guo, Y. B., (2003). An integral method to determine the mechanical behavior of materials in metal cutting. *Journal of Materials Processing Technology*, *142*(1), 72–81.

65. Gupta, M., & Wong, W. L. E., (2005). Enhancing overall mechanical performance of metallic materials using two-directional microwave assisted rapid sintering. *Scripta Materialia*, *52*(6), 479–483.

66. Hahn, G. T., & Rosenfield, A. R., (1975). Metallurgical factors affecting fracture toughness of aluminum alloys. *Metallurgical Transactions A*, *6*(4), 653–668.

67. Haynes, R., (1981). Reviews on the deformation behavior of materials. The mechanical behavior of sintered metals. *Freund Publishing House*, *1*(1), 101.

68. Henriksson, M., Berglund, L. A., Isaksson, P., Lindstrom, T., & Nishino, T., (2008). Cellulose nanopaper structures of high toughness. *Biomacromolecules*, *9*(6), 1579–1585.

69. Hjelmstad, K. D., (2007). *Fundamentals of Structural Mechanics* (Vol. 1, pp. 1–20). Springer Science & Business Media.

70. Hsiao, H. M., Daniel, I. M., & Cordes, R. D., (1999). Strain rate effects on the transverse compressive and shear behavior of unidirectional composites. *Journal of Composite Materials*, *33*(17), 1620–1642.

71. Hudson, J. A., Liu, E., & Crampin, S., (1996). The mechanical properties of materials with interconnected cracks and pores. *Geophysical Journal International, 124*(1), 105–112.

72. Hutchinson, J. W., (1968). Plastic stress and strain fields at a crack tip. *Journal of the Mechanics and Physics of Solids, 16*(5), 337–342.

73. Ibrahim, E. F., Price, E. G., & Wysiekierski, A. G., (1972). Creep and stress-rupture of high strength zirconium alloys. *Canadian Metallurgical Quarterly, 11*(1), 273–283.

74. Ilie, N., Bucuta, S., & Draenert, M., (2013). Bulk-fill resin-based composites: An *in vitro* assessment of their mechanical performance. *Operative Dentistry, 38*(6), 618–625.

75. Jakobsen, M., Hudson, J. A., Minshull, T. A., & Singh, S. C., (2000). Elastic properties of hydrate-bearing sediments using effective medium theory. *Journal of Geophysical Research: Solid Earth, 105*(B1), 561–577.

76. Jayaraman, K., & Bhattacharyya, D., (2004). Mechanical performance of woodfibre-waste plastic composite materials. *Resources, Conservation and Recycling, 41*(4), 307–319.

77. Jeffery, G. B., (1921). IX. Plane stress and plane strain in bipolar co-ordinates. philosophical transactions of the royal society of London. *Series A, Containing Papers of a Mathematical or Physical Character, 221*(582–593), 265–293.

78. Jones, D. B., Nolte, H., Scholübbers, J. G., Turner, E., & Veltel, D., (1991). Biochemical signal transduction of mechanical strain in osteoblast-like cells. *Biomaterials, 12*(2), 101–110.

79. Joun, M., Eom, J. G., & Lee, M. C., (2008). A new method for acquiring true stress–strain curves over a large range of strains using a tensile test and finite element method. *Mechanics of Materials, 40*(7), 586–593.

80. Kelly, A., & Tyson, A. W., (1965). Tensile properties of fiber-reinforced metals: Copper/tungsten and copper/molybdenum. *Journal of the Mechanics and Physics of Solids, 13*(6), 329–350.

81. Khanafer, K., Duprey, A., Schlicht, M., & Berguer, R., (2009). Effects of strain rate, mixing ratio, and stress-strain definition on the mechanical behavior of the polydimethylsiloxane (PDMS) material as related to its biological applications. *Biomedical Microdevices, 11*(2), 503.

82. Kim, B. S., Nikolovski, J., Bonadio, J., & Mooney, D. J., (1999). Cyclic mechanical strain regulates the development of engineered smooth

muscle tissue. *Nature Biotechnology*, *17*(10), 979.

83. Kim, W. H., & Laird, C., (1978). Crack nucleation and stage, I propagation in high strain fatigue—II. Mechanism. *Acta Metallurgica.*, *26*(5), 789–799.

84. Kolsky, H., (1949). An investigation of the mechanical properties of materials at very high rates of loading. *Proceedings of the Physical Society: Section B*, *62*(11), 676.

85. Komuro, I., & Yazaki, Y., (1993). Control of cardiac gene expression by mechanical stress. *Annual Review of Physiology*, *55*(1), 55–75.

86. Kontou, E., & Farasoglou, P., (1998). Determination of the true stress-strain behavior of polypropylene. *Journal of Materials Science*, *33*(1), 147–153.

87. Ku, H., Wang, H., Pattarachaiyakoop, N., & Trada, M., (2011). A review on the tensile properties of natural fiber reinforced polymer composites. *Composites Part B: Engineering*, *42*(4), 856–873.

88. Kukharev, A. V., (2015). *Some Stages of Evolution and Prospects of Tribo-Fatigue, 1*(1), 1–20.

89. Lade, P. V., & Duncan, J. M., (1975). Elastoplastic stress-strain theory for cohesionless soil. *Journal of Geotechnical and Geoenvironmental Engineering*, *101*(1167), 11–50.

90. Lawn, B. R., & Marshall, D. B., (1979). Hardness, toughness, and brittleness: An indentation analysis. *Journal of the American Ceramic Society*, *62*(7/8), 347–350.

91. Lee, J. E., Ahn, G., Shim, J., Lee, Y. S., & Ryu, S., (2012). Optical separation of mechanical strain from charge doping in graphene. *Nature Communications*, *3*, 1024.

92. Lehmann, B., Friedrich, K., Wu, C. L., Zhang, M. Q., & Rong, M. Z., (2003). Improvement of notch toughness of low nano-SiO_2 filled polypropylene composites. *Journal of Materials Science Letters*, *22*(14), 1027–1030.

93. Leis, B. N., Eiber, R. J., Carlson, L., & Gilroy-Scott, A., (1998). Relationship between apparent (total) Charpyvee-notch toughness and the corresponding dynamic crack-propagation resistance. In: *1998 2nd International Pipeline Conference* (Vol. 1, pp. 723–731). American Society of Mechanical Engineers Digital Collection.

94. LeRoux, M. A., Guilak, F., & Setton, L. A., (1999). Compressive and shear properties of alginate gel: Effects of sodium ions and alginate

concentration. *Journal of Biomedical Materials Research: An Official Journal of the Society for Biomaterials, the Japanese Society for Biomaterials, and the Australian Society for Biomaterials and the Korean Society for Biomaterials, 47*(1), 46–53.

95. Liao, J., Hotta, M., Kaneko, K., & Kondoh, K., (2009). Enhanced impact toughness of magnesium alloy by grain refinement. *Scripta Materialia, 61*(2), 208–211.

96. Lin, Y., Chen, H., Chan, C. M., & Wu, J., (2008). High impact toughness polypropylene/CaCO3 nanocomposites and the toughening mechanism. *Macromolecules, 41*(23), 9204–9213.

97. Ling, Y., (1996). Uniaxial true stress-strain after necking. *AMP Journal of Technology, 5*(1), 37–48.

98. Low, J. R., & Brown, W. F., (1981). *Fracture Toughness Testing and its Applications* (Vol. 381, pp. 11–33). ASTM International.

99. Lowhaphandu, P., & Lewandowski, J. J., (1998). Fracture toughness and notched toughness of bulk amorphous alloy: Zr-Ti-Ni-Cu-Be. *Scripta Materialia, 38*(12), 1–25.

100. Ma, A., Suzuki, K., Nishida, Y., Saito, N., Shigematsu, I., Takagi, M., & Imura, T., (2005). Impact toughness of an ultrafine-grained Al–11 mass% Si alloy processed by rotary-die equal-channel angular pressing. *Acta Materialia, 53*(1), 211–220.

101. MacKay, R. A., & Maier, R. D., (1982). The influence of orientation on the stress rupture properties of nickel-base superalloy single crystals. *Metallurgical Transactions A, 13*(10), 1747–1754.

102. Matsuishi, M., & Endo, T., (1968). Fatigue of metals subjected to varying stress. *Japan Society of Mechanical Engineers, Fukuoka, Japan, 68*(2), 37–40.

103. McClung, R. C., Thacker, B. H., & Roy, S., (1991). Finite element visualization of fatigue crack closure in plane stress and plane strain. *International Journal of Fracture, 50*(1), 27–49.

104. McKamey, C. G., Maziasz, P. J., & Jones, J. W., (1992). Effect of addition of molybdenum or niobium on creep-rupture properties of Fe 3 Al. *Journal of Materials Research, 7*(8), 2089–2106.

105. Mecklenburg, M., Schuchardt, A., Mishra, Y. K., Kaps, S., Adelung, R., Lotnyk, A., & Schulte, K., (2012). Aerographite: Ultra-lightweight, flexible nanowall, carbon microtube material with outstanding mechanical performance. *Advanced Materials, 24*(26), 3486–3490.

106. Monteiro, S. N., Terrones, L. A. H., & D'almeida, J. R. M., (2008). Mechanical performance of coir fiber/polyester composites. *Polymer Testing*, *27*(5), 591–595.

107. Murakami, Y., & Miller, K. J., (2005). What is fatigue damage? A view point from the observation of low cycle fatigue process. *International Journal of Fatigue*, *27*(8), 991–1005.

108. Murayama, T., (1978). *Dynamic Mechanical Analysis of Polymeric Material* (Vol. 1, p. 21). Amsterdam: Elsevier.

109. Murrell, S. A. F., (1964). The theory of the propagation of elliptical Griffith cracks under various conditions of plane strain or plane stress: Part I. *British Journal of Applied Physics*, *15*(10), 1195.

110. Nagendra, N., Ramamurty, U., Goh, T. T., & Li, Y., (2000). Effect of crystallinity on the impact toughness of a La-based bulk metallic glass. *Acta Materialia*, *48*(10), 2603–2615.

111. Nasser, M. S., & James, A. E., (2008). Compressive and shear properties of flocculated kaolinite-polyacrylamide suspensions. *Colloids and Surfaces A: Physicochemical and Engineering Aspects*, *317*(1–3), 211–221.

112. Nathal, M. V., Maier, R. D., & Ebert, L. J., (1982). The influence of cobalt on the tensile and stress-rupture properties of the nickel-base superalloy mar-m247. *Metallurgical Transactions A*, *13*(10), 1767–1774.

113. Neal, J. A., Mozhdehi, D., & Guan, Z., (2015). Enhancing mechanical performance of a covalent self-healing material by sacrificial noncovalent bonds. *Journal of the American Chemical Society*, *137*(14), 4846–4850.

114. Nomura, S., & Takano-Yamamoto, T., (2000). Molecular events caused by mechanical stress in bone. *Matrix Biology*, *19*(2), 91–96.

115. Norris Jr, D. M., (1979). Computer simulation of the Charpy V-notch toughness test. *Engineering Fracture Mechanics*, *11*(2), 261–274.

116. Oka, Y. I., & Yoshida, T., (2005). Practical estimation of erosion damage caused by solid particle impact: Part 2: Mechanical properties of materials directly associated with erosion damage. *Wear*, *259*(1–6), 102–109.

117. Oliver, J., (1996). Modeling strong discontinuities in solid mechanics via strain softening constitutive equations. Part 1: Fundamentals. *International Journal for Numerical Methods in Engineering*, *39*(21),

3575–3600.

118. Owan, I., Burr, D. B., Turner, C. H., Qiu, J., Tu, Y., Onyia, J. E., & Duncan, R. L., (1997). Mechanotransduction in bone: Osteoblasts are more responsive to fluid forces than mechanical strain. *American Journal of Physiology-Cell Physiology, 273*(3), C810–C815.

119. Palmgren, A., (1924). Die lebensdauer von kugellargern. *Zeitshrift des Vereines Duetsher Ingenieure, 68*(4), 339.

120. Plaut, M. A. R. S., Lichtenstein, L. M., & Henney, C. S., (1975). Properties of a subpopulation of T cells bearing histamine receptors. *The Journal of Clinical Investigation, 55*(4), 856–874.

121. Popescu, R., Deodatis, G., & Nobahar, A., (2005). Effects of random heterogeneity of soil properties on bearing capacity. *Probabilistic Engineering Mechanics, 20*(4), 324–341.

122. Queheillalt, D. T., Murty, Y., & Wadley, H. N., (2008). Mechanical properties of an extruded pyramidal lattice truss sandwich structure. *Scripta Materialia, 58*(1), 76–79.

123. Rankine, W. J. M., (1843). On the causes of the unexpected breakage of the journals of railway axles; and on the means of preventing such accidents by observing the law of continuity in their construction. In: *Minutes of the Proceedings of the Institution of Civil Engineers* (Vol. 2, No. 1843, pp. 105–107). Thomas Telford-ICE Virtual Library.

124. Rice, J., & Rosengren, G. F., (1968). Plane strain deformation near a crack tip in a power-law hardening material. *Journal of the Mechanics and Physics of Solids, 16*(1), 1–12.

125. Ritchie, R. O., (2011). The conflicts between strength and toughness. *Nature Materials, 10*(11), 817.

126. Ritchie, R. O., Knott, J. F., & Rice, J. R., (1973). On the relationship between critical tensile stress and fracture toughness in mild steel. *Journal of the Mechanics and Physics of Solids, 21*(6), 395–410.

127. Rubin, C. T., & Lanyon, L. E., (1985). Regulation of bone mass by mechanical strain magnitude. *Calcified Tissue International, 37*(4), 411–417.

128. Runesson, K., Ottosen, N. S., & Dunja, P., (1991). Discontinuous bifurcations of elastic-plastic solutions at plane stress and plane strain. *International Journal of Plasticity, 7*(1/2), 99–121.

129. Sadoshima, J., & Izumo, S., (1997). The cellular and molecular response of cardiac myocytes to mechanical stress. *Annual Review of*

Physiology, *59*(1), 551–571.

130. Sánchez-Arévalo, F. M., & Pulos, G., (2008). Use of digital image correlation to determine the mechanical behavior of materials. *Materials Characterization*, *59*(11), 1572–1579.

131. Schijve, J., (1978). Internal fatigue cracks are growing in vacuum. *Engineering Fracture Mechanics*, *10*(2), 359–370.

132. Schijve, J., (2003). Fatigue of structures and materials in the 20th century and the state of the art. *International Journal of Fatigue*, *25*(8), 679–702.

133. Schütz, W., (1996). A history of fatigue. *Engineering Fracture Mechanics*, *54*(2), 263–300.

134. Servi, I. S., & Grant, N. J., (1951). Creep and stress rupture behavior of aluminum as a function of purity. *JOM*, *3*(10), 909–916.

135. Shen, Y. F., Lu, L., Lu, Q. H., Jin, Z. H., & Lu, K., (2005). Tensile properties of copper with nano-scale twins. *Scripta Materialia*, *52*(10), 989–994.

136. Sherby, O. D., & Burke, P. M., (1968). Mechanical behavior of crystalline solids at elevated temperature. *Progress in Materials Science*, *13*, 323–390.

137. Shigley, J. E., Mischke, C. R., & Budynas, R. G., (2007). *Mechanical Engineering Design* (Vol. 4, pp. 1–20).

138. Skaggs, D. L., Weidenbaum, M., Iatridis, J. C., Ratcliffe, A., & Mow, V. C., (1994). Regional variation in tensile properties and biochemical composition of the human lumbar anulus fibrosus. *Spine, 19*(12), 1310–1319.

139. Song, Y. Y., Ping, D. H., Yin, F. X., Li, X. Y., & Li, Y. Y., (2010). Microstructural evolution and low temperature impact toughness of a Fe–13% Cr–4% Ni–Mo martensitic stainless steel. *Materials Science and Engineering: A, 527*(3), 614–618.

140. Sosnovskiy, L. A., (2003). Fundamentals of tribo-fatigue. *BelGUT, Gomel*, *1*, 246.

141. Stephens, R. I., Fatemi, A., Stephens, R. R., & Fuchs, H. O., (2000). *Metal Fatigue in Engineering* (Vol. 1, pp. 4–46). John Wiley & Sons.

142. Stolyarov, V. V., Valiev, R. Z., & Zhu, Y. T., (2006). Enhanced low-temperature impact toughness of nanostructured Ti. *Applied Physics Letters*, *88*(4), 041905.

143. Sugimura, Y., Meyer, J., He, M. Y., Bart-Smith, H., Grenstedt, J., & Evans, A. G., (1997). On the mechanical performance of closed cell Al alloy foams. *Acta Materialia, 45*(12), 5245–5259.

144. Sun, W. R., Guo, S. R., Lu, D. Z., & Hu, Z. Q., (1997). Effect of phosphorus on the microstructure and stress rupture properties in an Fe-Ni-Cr base superalloy. *Metallurgical and Materials Transactions A, 28*(3), 649–654.

145. Svensson, L. E., & Gretoft, B., (1990). Microstructure and impact toughness of C--Mn weld metals. *Welding Journal, 69*(12), 454.

146. Tan, H., Li, L., Chen, Z., Song, Y., & Zheng, Q., (2005). Phase morphology and impact toughness of impact polypropylene copolymer. *Polymer, 46*(10), 3522–3527.

147. Tanaka, E., Kawai, N., Hanaoka, K., van Eijden, T. M. G. J., Sasaki, A., Aoyama, J., & Tanne, K., (2004). Shear properties of the temporomandibular joint disc in relation to compressive and shear strain. *Journal of Dental Research, 83*(6), 476–479.

148. Thomas, G. B., & Gibbons, T. B., (1979). Influence of trace elements on creep and stress-rupture properties of Nimonic 105. *Metals Technology, 6*(1), 95–101.

149. Thompson, M. S., McCarthy, I. D., Lidgren, L., & Ryd, L., (2003). Compressive and shear properties of commercially available polyurethane foams. *J. Biomech. Eng., 125*(5), 732–734.

150. Titze, I. R., (1994). Mechanical stress in phonation. *Journal of Voice, 8*(2), 99–105.

151. Van Dick, R., & Wagner, U., (2001). Stress and strain in teaching: A structural equation approach. *British Journal of Educational Psychology, 71*(2), 243–259.

152. Vegas, M. R., & Del Yerro, J. L. M., (2013). Stiffness, compliance, resilience, and creep deformation: Understanding implant-soft tissue dynamics in the augmented breast: Fundamentals based on materials science. *Aesthetic Plastic Surgery, 37*(5), 922–930.

153. Vladimir, M. P., & Imanaka, T., (1997). Legislation and research activity in Belarus about the radiological consequences of the Chernobyl accident: Historical review and present situation. *Japanese Journal of Health Physics, 32*(1), 81–96.

154. Waite, W. F., Santamarina, J. C., Cortes, D. D., Dugan, B., Espinoza, D. N., Germaine, J., & Soga, K., (2009). Physical properties of hydrate-bearing sediments. *Reviews of Geophysics, 47*(4), 1–22.

155. Weibull, W., (1952). The statistical aspect of fatigue failure and its consequences. In: *Fatigue and Fracture of Metals; Massachusetts Institute of Technology* (Vol. 1, No. 1, pp. 182–196). John Wiley & Sons: New York, NY, USA.

156. Weibull, W., (2013). *Fatigue Testing and Analysis of Results* (Vol. 1, pp. 2–26). Elsevier.

157. Wesseling, P., Nieh, T. G., Wang, W. H., & Lewandowski, J. J., (2004). Preliminary assessment of flow, notch toughness, and high temperature behavior of Cu60Zr20Hf10Ti10 bulk metallic glass. *Scripta Materialia, 51*(2), 151–154.

158. Westman, M., (2001). Stress and strain crossover. *Human Relations, 54*(6), 717–751.

159. Wolf, E., (1970). Fatigue crack closure under cyclic tension. *Engineering Fracture Mechanics, 2*(1), 37–45.

160. Wong, E. W., Sheehan, P. E., & Lieber, C. M., (1997). Nanobeam mechanics: Elasticity, strength, and toughness of nanorods and nanotubes. *Science, 277*(5334), 1971–1975.

161. Xu, C. N., Watanabe, T., Akiyama, M., & Zheng, X. G., (1999). Artificial skin to sense mechanical stress by visible light emission. *Applied Physics Letters, 74*(9), 1236–1238.

162. Yan, W., Shan, Y. Y., & Yang, K., (2006). Effect of TiN inclusions on the impact toughness of low-carbon micro alloyed steels. *Metallurgical and Materials Transactions A, 37*(7), 2147–2158.

163. Yang, F. A. C. M., Chong, A. C. M., Lam, D. C. C., & Tong, P., (2002). Couple stress-based strain gradient theory for elasticity. *International Journal of Solids and Structures, 39*(10), 2731–2743.

164. Yun, H. M., & DiCarlo, J. A., (1999). Comparison of the tensile, creep, and rupture strength properties of stoichiometric SiC fibers. In: *Proceedings of the 23rd Annual Conference on Composites, Materials and Structures* (Vol. 20, pp. 259–72).

165. Zhang, P., Ma, L., Fan, F., Zeng, Z., Peng, C., Loya, P. E., & Ajayan, P. M., (2014). Fracture toughness of graphene. *Nature Communications, 5*, 3782.

166. Zhu, F., Bai, P., Zhang, J., Lei, D., & He, X., (2015). Measurement of true stress–strain curves and evolution of plastic zone of low carbon steel under uniaxial tension using digital image correlation. *Optics and Lasers in Engineering, 65*, 81–88.

Chapter

8

Heat Transfer by Conduction, Convection, and Radiation

CONTENTS

8.1. INTRODUCTION

It is known that heat is the energy associated to the motion of molecules. Roughly, molecules having higher heat energy will move faster and molecules having less heat energy will move slower. It is also known that as the molecules heat up and start moving faster, the molecules spread apart, and the bodies expand. This is known as thermal expansion (Heijnen and Van't Riet, 1984; Chang and You, 1997) (Figure 8.1).

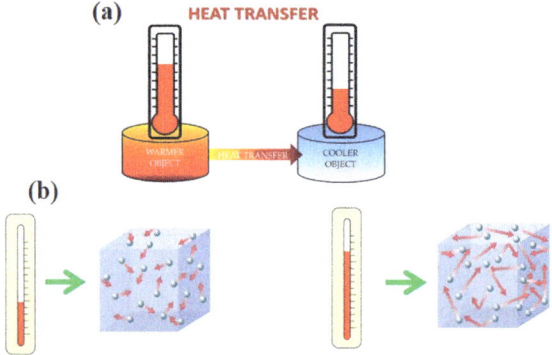

Figure 8.1. (a) Heat transfer from the warmer object to cooler one. (b) The temperature of cooler object increases slowly.

Heat is always considered moving. If two substances or objects are taken which are at different temperatures, the heat will move out of warmer substance or objects, and into the cooler substance or object. This transfer of heat will continue until the objects or substances are the same temperatures (Kou, 1996; Kalb and Seader, 1972).

So how does exactly heat move out of the object and into another object? This is known as heat transfer. Heat can transfer in three ways:

- Conduction;
- Convection;
- Radiation.

As these three types of heat transfer will be discussed, the attention will be paid to:

- Through which object the heat is moving (gases, liquids, and solids, or the empty space);
- How is the heat being transferred (currents, touch or waves).

Thermodynamics usually makes reference to heat transfer between the systems. Frequently these laws do not satisfactorily describe the processes of heat transfer, so more accurate rules must be introduced to explain what happens (Peng and Wang, 1993; Kirillov et al., 1995). The control of the heat transfer is vital to study in order to design the suitable tools to transfer the thermal energy from a medium to the other medium. This module presents the heat transfer and transport laws of conduction, convection, and radiation. The laws presented include Fourier's law, Stefan-Boltzmann law and Newton's law of cooling. Other topics that are described include Biot numbers, 1-D heat diffusion equation, and Wein's law. This aid as an outline to complicated nature of the thermal energy transfer (Sass, 1967; Lee, 1983) (Figure 8.2).

Figure 8.2. Heat transfer from the hot sand to human feet.

8.2. APPROACHES OF HEAT TRANSFER

When the difference of temperature exists, heat will generally flow from the hot to cold. Heat can transfer amongst the two mediums by conduction, radiation, and convection whenever a temperature difference exists. Remember the first law of thermodynamics (Arpaci and Arpaci, 1966; Özişik et al., 2017). The rate that the heat will transfer in the closed system is given in the form below:

$$Q = W + \frac{dU}{dt}$$

(1)

where:

$Q \rightarrow$ heat transfer rate.

$W \rightarrow$ work transfer rate.

dU/dt → net change in the total energy of the system.

Normally, the transfer of heat can be examined without work being involved (Hauf and Grigull, 1970; Ganji and Rajabi, 2006). However, the real systems can comprise work in their analysis. In the situation of only *p*, *the dv* work happening, Equation (1) becomes:

$$Q = p.\frac{dV}{dt} + \frac{dU}{dt} \qquad (2)$$

With two special circumstances: constant volume and constant pressure. In the case of the constant volume:

$$Q = \frac{dU}{dt} = mc_v.\frac{dT}{dt} \qquad (3)$$

Specific heat capacity is c_v. For the constant pressure:

$$Q = \frac{dH}{dt} = mc_P.\frac{dT}{dt} \qquad (4)$$

With enthalpy H = U+ p.V and c_p is the specific heat capacity. The specific heat capacities will generally be equal in the incompressible liquid, with the constant volume at any pressure, interpreting $c_v = c_p = c$. The rate of heat transfer becomes:

$$Q = \frac{dU}{dt} = m.c.\frac{dT}{dt} \qquad (5)$$

It is not always known directly, so most of the time it can't be utilized to find Q. To accomplish this, one must utilize the transport laws to precisely forecast the rate of heat transfer (Aziz and Na, 1984; Rohsenow et al., 1985). These laws are the Fourier's law, Stefan-Boltzmann law and Newton's law of cooling introduced in following sections.

8.3. CONDUCTION

Conduction is how the heat transfers through the direct contact with the objects that are touching. Any time that 2 substances or objects touch, the hotter substance passes heat to a cooler substance (Kreith and Black, 1980; Incropera et al., 2007).

Let's think of a row of the dominoes that are lined up. When the first domino is pushed, it usually bumps into the second one, the second bumps into the third, all way down the line. Heat conduction is just like dominoes.

Imagine that one end of the metal pole is placed into the fire. The molecules on fire end will normally get hot. Each of the hot molecules will transfer the heat to the next molecule, which will transfer the heat to next molecule and so on. Before it is known, the heat has moved all the way to the end of the metal pole (Jaeger and Carslaw, 1959; Bourdi et al., 1983) (Figure 8.3).

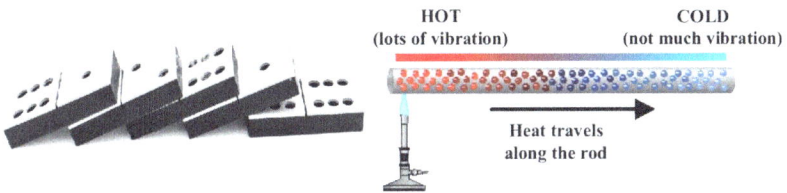

Figure 8.3. Heat conduction in the metallic pole.

Source: https://www.slideshare.net/venxaiimada/heat-transfer-15633227.

Some materials are far better conductors than the others. That is because some of the materials pass heat more easily. Metals are the great conductors. That is why metal substances get hot easily. Wood and plastic are poor conductors. They will get hot, but it will take longer time for these objects to pass the heat energy from one molecule to another (Biot, 1970; Bejan and Kraus, 2003).

Similarly, solids are much better conductors than gases or liquids. That is because solids possess molecules that are tightly packed together, so it is much easier for these molecules to pass the heat. The molecules in gases and liquids are spread quite further apart, so they are not touching as much. It takes longer for gases and liquids to conduct heat (VanSant, 1983; Kurzweg, 1985) (Figure 8.4).

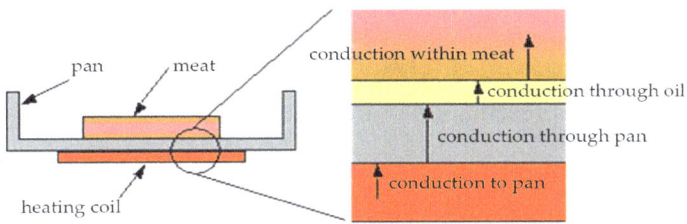

Figure 8.4. Conduction phenomena in the cooking operation.

Source: https://thermtest.com/how-understanding-thermal-conductivity-will-make-you-a-better-cook.

There are several examples of the process of heat conduction. Any time two substances or objects touch, heat conduction will take place. Touching the iron which is hot is an example of heat conduction—the heat flows from the iron and into the hand. So is holding the cube of ice—the heat passes out of the hand and into an ice cube.

8.3.1. Heat Flux and the Thermal Conductivity

Conduction is a transfer of thermal energy through the contact of particles. Small particles transfer potential and kinetic energy as they strike and vibrate with the other particles. Two materials can share energy only by conduction if these two materials are in indirect or direct contact with one another. The rate of flow of this heat energy is called the heat flux (Xuan and Li, 2000; You et al., 2003).

Heat flux or the thermal flux is defined as the measurement of rate of heat transfer per unit of the area, expressed in W/m^2 (watts per square meter). Mathematically, it is the vector quantity, its magnitude is characterized as q.

$q = Q/A$ (6)

here:

$Q \rightarrow$ heat transfer rate.

$A \rightarrow$ cross-sectional area.

Heat flux from the thermal conduction is proportional to a temperature gradient across the object and is opposite in polarity. It changes by the constant k, **thermal conductivity** of the material. The units of thermal conductivity are watts per meter Kelvin (W/m.K). It is dependent on material and can be determined experimentally. This relationship is called Fourier's law of heat transfer (Berber et al., 2000; Zhu and Miller, 2000).

$$q = -k.\frac{dT}{dx}$$ (7)

This is a 1-D representation of the heat flux (Figure 8.5).

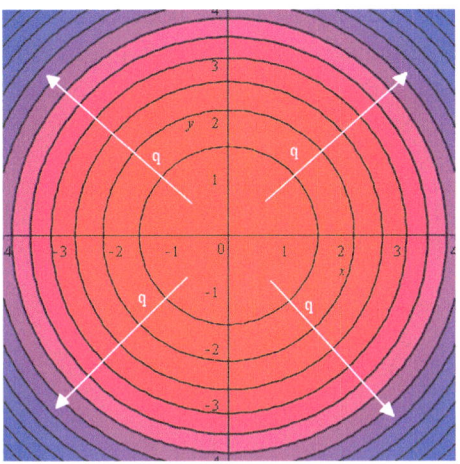

Figure 8.5. Heat flux displayed on the temperature distribution graph.

Source: https://www.maplesoft.com/content/EngineeringFundamentals/51/Ma-pleDocument_49/heat_transfer.pdf.

As temperature flows from the hot to cold, heat flux will be positive (+ve) if the rate of change of temperature gradient decreases. In multidimensional representation:

$$q = -k\Delta T = -k \left(\frac{\partial T}{\partial x} i + \frac{\partial T}{\partial y} j + \frac{\partial T}{\partial z} k \right) \tag{8}$$

It is at times more suitable to work with the scalar form of the above equation, such as one-dimensional problems where direction of the heat flow is determined easily.

Recalling that heat flows from the hot to cold, the heat flux can be determined by:

$$q = \frac{k\Delta T}{L} \tag{9}$$

where:

$L \rightarrow$ thickness of material in the direction of heat flow.

k & $T \rightarrow$ positive quantities.

Different materials will normally conduct heat better than others. Copper, for instance, has very high thermal conductivity since it is good electrical conductor and is able to move electrons and transfer energy easily.

On the contrary, diamond, which is the poor electrical conductor, usually transfers heat even better as compared to copper because of its effective lattice structure (Narayana and Sato, 2012). Gases are far better insulators as the molecules possess more space in order to move around and do not interact as well as the solids.

Materials with the high thermal conductivity have a smaller temperature gradient as (from Equation 8),

$$-\frac{dT}{dx} = \frac{q}{k}$$

(10)

or more usually:

$$\frac{dT}{dx} \propto \frac{q1}{k}$$

(11)

For the larger thermal conductivities, the temperature gradient will vary less.

8.3.2. One-Dimensional Heat Diffusion Equation

In the earlier example, the temperature distribution was obtained by equating heat flux through each material (Figure 8.6). This can be done once heat flux reaches a steady state (Xuan and Li, 2003; Hayat et al., 2016). What can be done if heat flux is not constant?

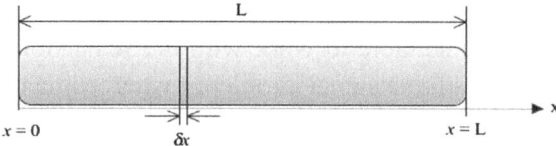

Figure 8.6. A uniform rod having length L.

Source:https://www.maplesoft.com/content/EngineeringFundamentals/51/Ma-pleDocument_49/heat_transfer.pdf.

Consider the metal rod (Figure 8.6) having insulated sides and the ends exposed. This rod has the length L that lengthens from x = (0 to L). This is the uniform rod so suppose the specific heat c, thermal conductivity k, density ϱ and the cross-sectional area A are constant. The temperature T(x, t) varies with both time t and position x (Zhu et al., 2001). The amount of the heat in the rod at time is given as,

$$U(t) = \rho \cdot c \cdot T(x, T) \cdot \delta x \qquad (12)$$

where:

$\varrho \rightarrow$ mass per unit length.

$\delta x \rightarrow$ differential length of the rod.

   After δ seconds, the amount of the heat in rod will be:

$$U(t + \delta t) = \rho \cdot c \cdot T(x, t + \delta t) \cdot \delta x$$

   The change in the heat will be the difference between these.

$$U(t + \delta t) - U(t) = \rho \cdot c \cdot (T(x, t + \delta t) - T(x, t)) \cdot \delta x \qquad (13)$$

This should be equal to heat flowing into the rod at position x minus heat flowing out of rod at x + δx, for the same duration of the time δt. Recall from the Equation (7) that heat flow is proportional to a temperature gradient.

$$U(t + \delta t) - U(t) = \left[\left(-k \cdot \frac{\partial T}{\partial x} \right)_x - \left(-k \cdot \frac{\partial T}{\partial x} \right)_{x+\delta x} \right] \cdot \delta t \qquad (14)$$

   Equate these 2 terms and then divide by δx & δt.

$$\rho \cdot c \cdot \frac{(T(x, t + \delta t) - T(x, t))}{\delta t} = k \cdot \frac{\left(\frac{\partial T}{\partial x} \right)_{x+\delta x} - \left(\frac{\partial T}{\partial x} \right)_x}{\delta t} \qquad (15)$$

   By taking the limit as $\delta x \rightarrow 0$ * $\delta t \rightarrow 0$, 1-dimensional heat diffusion equation is obtained.

$$\frac{\partial T}{\partial t} = \frac{k}{\rho \cdot c} \cdot \frac{\partial^2 T}{\partial x^2} \qquad (16)$$

   This can be shortened with the variable of thermal diffusivity $\alpha = \frac{k}{p.c}$ in $(\frac{m^2}{s})$

$$\frac{\partial T}{\partial t} = \frac{\alpha \partial^2 T}{\partial^2 x} \qquad (17)$$

8.4. CONVECTION

Convection is the process of how heat passes through the fluids. A fluid is something that has lightly moving molecules that can move from one place to the other easily. Gases and liquids are fluids (Eckert and Drake Jr, 1987).

One vital property of the fluids is that when heated they rise. That is as the molecules generally spread out and then move apart when they are heated. The hot fluid then becomes less thick and rises. The cooler fluid is less thick and thus it sinks down. This up and down motion produces what is called the convection currents. Convection currents are the circular movements of the heated fluids that aid to circulate the heat (Kakaç and Pramuanjaroenkij, 2009; Bejan, 2013) (Figure 8.7).

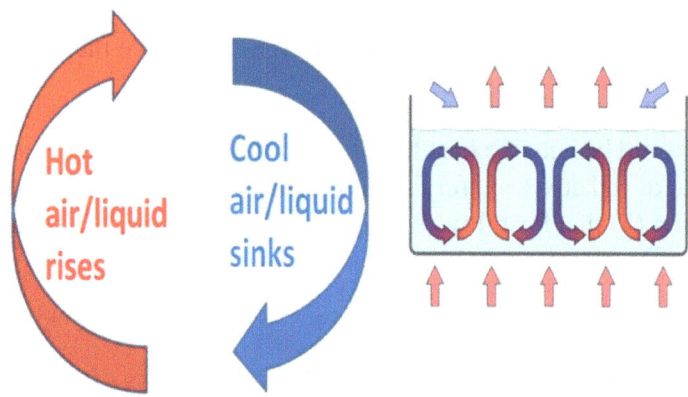

Figure 8.7. A typical illustration of the convection phenomenon.

Source: https://www.maplesoft.com/content/EngineeringFundamentals/51/MapleDocument_49/heat_transfer.pdf.

An everyday life example of the convection is the dinner soup, as is heated, first, the soup might be cold in a pan, the soup at bottom of the pan should have been nearest to hot stove burner, so, the soup at bottom is heated up first. As the soup is heated, molecules are spread apart and became less thick. So, the soup which is heated rises to the top (Morgan, 1975; Gee and Webb, 1980).

As the hot soup rose, the soup which is cooler at the top moves to the bottom.. As the soup is heated continuously, the cold soup sank, and the hot soup rose. Convection currents clarify why the air is cooler at the bottom of the room and hotter at the top. Convection currents also clarify why water is colder at deeper down the ocean, and hotter at the top (Fujii and Imura, 1972; Incropera et al., 2007).

One natural example of the convection currents is wind. As Sun shines down on the area of land, it generally heats the air over the ground. That warm air then rises. As the warm air rises, the cooler air travels in order to take place at the bottom. This moving air creates the winds. Wind happens all over the Earth since Earth heats unevenly. There are always warmer parts and colder parts. The wind blows from cooler parts of the Earth to warmer parts (Kuehn and Goldstein, 1976) (Figure 8.8).

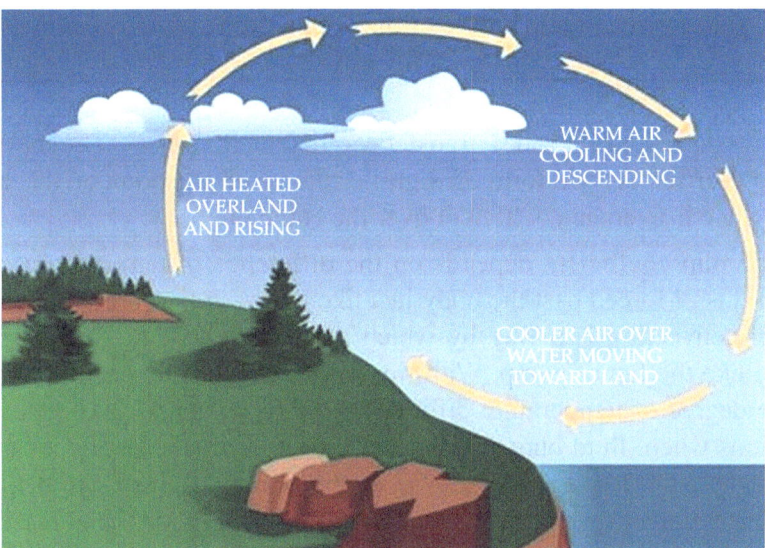

Figure 8.8. Example of the convection in winds.

Source: http://www.dauntless-soft.com/PRODUCTS/Freebies/Library/books/ AK/5-3.htm.

Other examples of the convection are: boiling the pot of water on the stove; using the hot radiator in order to warm air in the room; and utilizing the heated air to make the hot air balloon rise up into sky. Convection takes place when the fluid or the gas flows around the object. A small layer of the fluid establishes around the body, known as the boundary layer, where the heat diffuses from the object to the fluid. The thermal energy is carried away from object by the fluid (Wang and Peng, 1994).

Newton's law of the cooling describes that the temperature difference amongst the oncoming fluid and body is proportional to heat flow from the body. The steady-state equation of law of the cooling is given as:

$$q = \bar{h}.\left(T_{Body} - T_{Fluid}\right) \qquad (18)$$

where:

q →heat flow rate.

T_{Body} →temperature of body.

T_{Fluid} →the constant temperature of oncoming fluid.

h is film coefficient or the heat transfer coefficient, in. $\left(\dfrac{W}{m^2 k} \right)$ The coefficient of heat transfer has two forms. h signifies the value at a point on the surface while the $\overline{h}$ is average coefficient over the complete body.

The film coefficient depends on the difference of temperature only if the fluid isn't forced past the body just like it is with the forced convection. Forced convection is a way by which a fluid "flows" due to an external source like the fan or pump. With free or natural convection, h can change by the degree of temperature difference like $\Delta T^{1/2}$ and ΔT^3. This occurs in situations where fluid buoys up around the body or when body is very hot to boil the fluid, for instance. However, the examples given will be limited to the situations in which Newton's law of cooling relates or is reasonably precise (Figure 8.9).

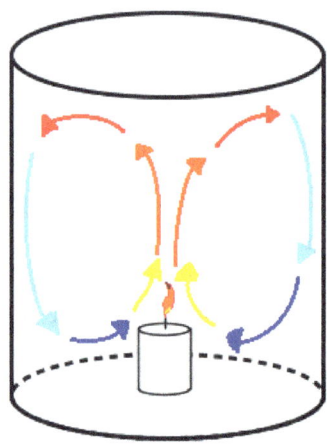

Figure 8.9. Natural convection. The hot air from flame magnifies and rises above the thicker cold air.

To forecast transient cooling of the convectively cooled objects, a lumped- capacity solution is used. Begin by remembering the first law statement:

$$Q = \frac{dU}{dt}$$

By replacing in equations for Q from the Equation (18) and $\frac{dU}{dt}$ from Equation (5), this will become:

$$-\bar{h}.A.(T - T_\infty) = \rho.c.V.\frac{d}{dt}(T - T_{Ref})$$

(19)

Although T_∞ signifies the temperature of convection fluid and T_{ref} is the temperature of the U defined at 0, derivatives of these two will both be zero as they are constants. The differences can be considered as equivalent (assume $T_{ref} = T_\infty$). The solution becomes as:

$$\ln(T - T_\infty) = -\frac{t}{\left(\dfrac{\rho.c.V}{\bar{h}.A}\right)} + C$$

(20)

The constant C can be determined at time $t = 0$.

$$C = \ln(T_i - T_\infty)$$

(21)

Replacing this in and then rearranging solves for cooling of the body. This is a lumped- capacity solution.

$$\frac{T - T_\infty}{T_i - T_\infty} = e^{-\frac{t}{\tau}}$$

(22)

where τ . is time constant, which binds together the material properties in the Equation (20), given by:

$$\tau = \rho.c.\frac{V}{\bar{h}.A}$$

(23)

The thermal conductivity of a material is not restricted to be constant. It is supposed that heating all the way through the object is uniform. In these circumstances, Biot number for the body is generally less than 0.1. For other cases, more parameters should be considered.

8.5. BIOT NUMBER

The Biot number is generally defined as:

The ratio amongst the rate that the body cools through conduction compared to the convective cooling (Boddington et al., 1983; Carrigan, 1988; Chen et al., 2001). It is a unitless quantity determined by equating the Fourier's law, Equation (8), to Newton's law of the cooling, Equation (18).

$$B_i = \frac{\overline{h}L}{k} \qquad (24)$$

aking the ratio among the temperature differences discloses the Biot number. The Figure 8.10 exhibits two different materials that are cooled by the convection (Hensen and Nakhi, 1994; Dincer and Hussain, 2004; Chen and Peng, 2005). The temperature across both materials is shown.

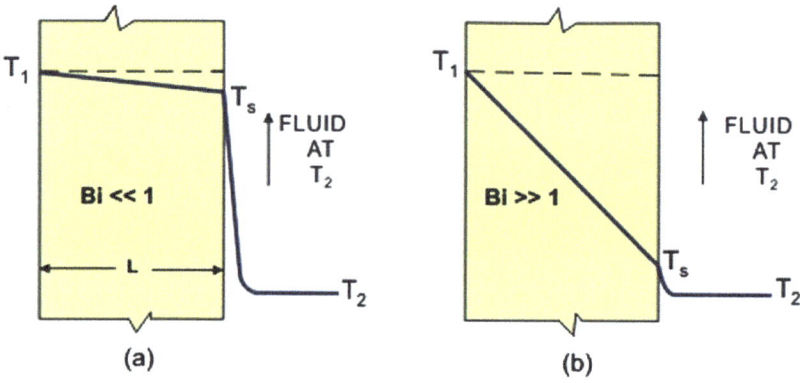

(a)   (b)

Figure 8.10. Low and high Biot numbers.

Source: http://www.thermopedia.com/content/585/.

In Figure 8.10 left object is considered to have a low Biot number. It has a low-temperature drop within body because of the larger thermal conductivity which is opposite to the second body having low conductivity (Pham and Willix, 1990; Huang and Yeh, 2002). Evidently, the Biot number must be anticipated below as there is larger drop amongst the surface temperature and ambient fluid temperature (T_s-T_∞). Rearrange Equation (24) to notice this explicitly (Xu et al., 2012; Bég et al., 2014).

$$Bi = \frac{\overline{h}.L}{k} = \frac{(T_1 - T_s)}{(T_s - T_\infty)} \qquad (25)$$

8.6. RADIATION

It is understood now that the conduction moves heat at ease through the solids and convection moves heat at ease through gases and liquids. So how does heat from the Sun comes down to the Earth? There are not any molecules in the space and how does one feel the heat from the campfire, even if one is sitting several feet away?

The answer to this is radiation. Radiation is how the heat moves through the places where there are not any molecules. Radiation is a form of electromagnetic energy. Recall it is learned that the electromagnetic energy moves in the waves? Well, radiation is the heat moving in waves. Radiation does not need molecules in order to allow the energy along (Howell et al., 2015; Sparrow, 2018).

All of the objects radiate heat but some of them radiate more heat as compared to the others. The biggest source of the radiation is Sun—it sends quite a huge amount of the heat to Earth through the electromagnetic waves (Viskanta and Mengüç, 1987) (Figure 8.11).

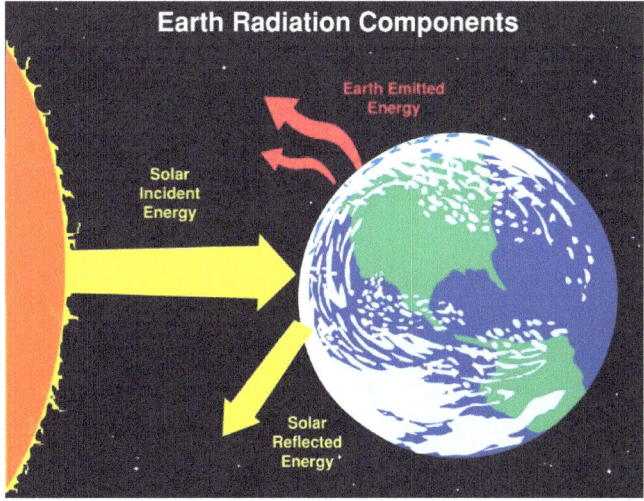

Figure 8.11. Radiation of sun coming towards the earth.

Source: https://visibleearth.nasa.gov/images/54217/earth-radiation-compo-nents.

Light bulbs also radiate heat. It can be tried by holding the hand a few inches away from the light bulb. The heat can be felt. In fact, an easy way to remember the radiation is that how one can feel the heat without touching it. Heat travels or moves through the empty space before reaching our hand. That's radiation. Fire is also an example of radiation. Even humans beings are an example. Our body radiates heat (Chai et al., 1994) (Figure 8.12).

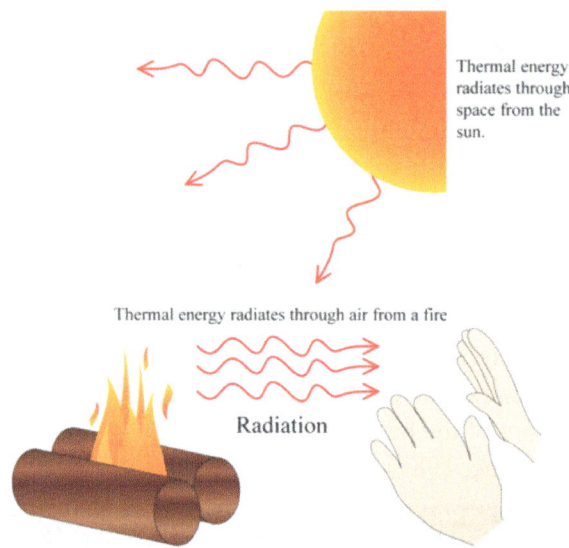

Figure 8.12. Heat radiation moves away from the fire-source.

Source: https://www.ck12.org/physics/thermal-radiation/lesson/Thermal-Radiation-MS-PS/.

All bodies continuously emit some kind of thermal energy by radiating heat, and energy can move between bodies in the form of the radiation. Thermal energy can radiate across the range of wavelengths but normally it is near to that of the infrared. Frequently, energy emitted by the radiation can be ignored in the existence of convection and conduction at low temperatures. However, at very high temperatures, radiation needs to be considered as the energy emission from the body changes as the 4th degree of absolute temperature (Edwards and Balakrishnan, 1973; Dombrovsky, 1996).

As the radiation hits an object, some energy might be absorbed, pass through the object, or reflect off from the surface. A black body is an object that does not reflect any radiation or let the energy pass through. It absorbs all of the incident radiation and then re-emits thermal energy at the rate dependent on the black body, not on the incident radiation which heats it (Dehghan and Behnia, 1996; Fu and Zhang, 2006) (Figure 8.13).

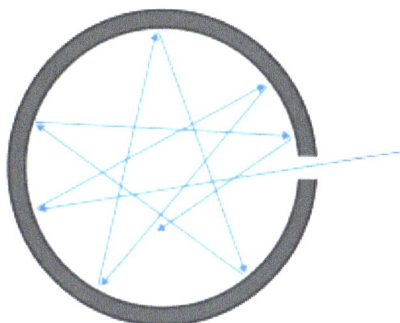

Figure 8.13. A black body as the cavity having a hole.

Source: https://en.wikipedia.org/wiki/Black_body.

A famous model of the black body is that of the cavity having a hole through which the radiation enters, as exhibited in Figure 8.13. The walls of cavity excellently absorb radiation and the surface of body emits the thermal energy. The rate at which the body emits energy reaches the theoretical maximum, given by Stefan-Boltzmann law (Cess, 1966).

Stefan-Boltzmann's law only applies to black bodies. It describes the directly proportional association amongst the energy emitted from unit area in 1 second, $e(T)$, to 4th power of the temperature.

$$e(T) = \sigma . T^4 \qquad (26)$$

where:

σ →Stefan-Boltzmann constant $\sigma = 5.6704 \cdot 10^{-8} \dfrac{[\![W]\!]}{[\![m^2]\!][\![K^4]\!]}$.

Now here, $e(T)$ might be mentioned as the irradiance, heat flux density, or emissive power and is given in watts per square meter. Stefan through experimentation. formulated this in 1879.

The radiation that the black body will give off is proportional to the temperature of the body. This radiation is generally given off as the distribution of energy across numerous wavelengths. For instance, an iron left in the fire will glow the dull red, emitting the energy mostly through the infrared light and some energy in visible spectrum. If it is to be heated to the hotter temperature, it would normally emit more visible light. Wein's law states that hotter an object, the shorter the wavelength an object will emit through the radiation. Wein's law also describes that peak wavelength that the radiating energy is emitted at is also proportional to its temperature.

$$\lambda_{max}.T = 2.898.10^{-3} \parallel m \parallel \llbracket K \rrbracket \qquad (27)$$

The constant of proportionality which is used is mentioned as the Wein's displacement constant. It must be observed that this relationship does not correspond with the peak frequency of radiation through $v = f\lambda$, which is the peak f_{max} and peak wavelength λ_{max} are not directly associated with each other. Regulate the temperature gauge in the Figure 8.14 and notice its effect on wavelength's distribution.

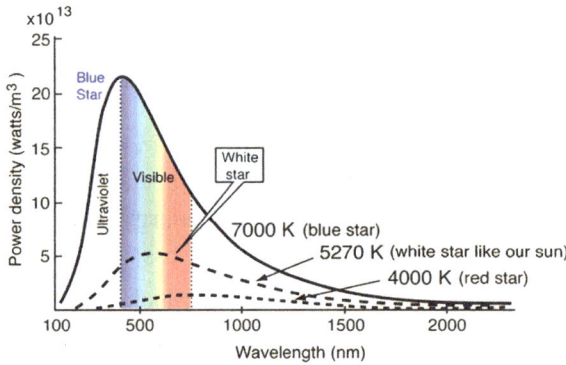

λ**Figure 8.14.** Emissive power versus wavelength chart with the adjustable temperature.

Source: http://hyperphysics.phy-astr.gsu.edu/hbase/wien.html.

The Figure 8.14 exhibits the wavelength distribution of the thermal radiation versus emissive power for the black body and shows the range of the visible color emitted by the body.

The temperature gauge also sets the surface temperature of the black body. The emissive power which depends on lambda was offered in 1901 by Max Planck and is described to be:

$$e_{\lambda} = \frac{2.\pi.h.c_0^2}{\lambda^5.\exp\left(\left(\dfrac{h.c_0}{k_B.T.\lambda}\right)-1\right)} \qquad (28)$$

where:

h →Planck's constant, $h = 6.62607004 \times 10$–34 m2 kg/s.

c_0 →speed of light.

k_B →Boltzmann's constant.

The net energy transferred by an object through radiation Q_{net} is the difference amongst $Q_{1\to2}$. If any object radiated heat towards the other object, supposing both of the objects are black bodies, then net heat transferred is given as:

$$Q_{net} = \sigma \cdot A_1 \cdot (T_1^4 - T_2^4) = A_1 \cdot [e_1(T) - e_2(T)] \qquad (29)$$

This net transfer supposes that all of the radiating energy from object one *sees* the second object. In situations where there are many objects, or some of the radiation diffuses away, this net transfer is multiplied by the view factor. This is a fraction of the energy intercepted by the object 2 from the object 1.

$$Q_{net} = F_{1\to2} \cdot \sigma \cdot A_1 \cdot (T_1^4 - T_2^4) \qquad (30)$$

If the objects aren't perfect black bodies then this view factor is substituted by transfer factor which accounts for different geometries and surface properties of the objects.

8.7. COMBINED HEAT TRANSFER

Heat passes from warmer object to the cooler object 'til all the objects are the same temperature. Conduction is how the heat travels among the objects that are touching. Conduction travels quicker through the solids, but gases and liquids can also conduct heat (Rousse, 2000; Hsiao, 2017). Some materials such as metal can rapidly conduct heat, whereas other materials conduct heat gradually (Chung and Kim, 1984; Yoshida et al., 1990).

Convection is how the heat travels through the fluids-gases, and liquids. Hot fluids rise, whereas cold fluids generally sink down (Tan and Howell, 1991; Dehghan and Behnia, 1996; Mezrhab et al., 2006). This up and down motion is known as convection current. Convection current distributes the heat in a circular, up, and down pattern. Radiation is how the heat travels through the empty space. Radiation does not need molecules to travel through. Any time one feels the heat without touching it, he/she is experiencing radiation (Balaji and Venkateshan, 1995; Reddy and Kumar, 2008) (Figure 8.15).

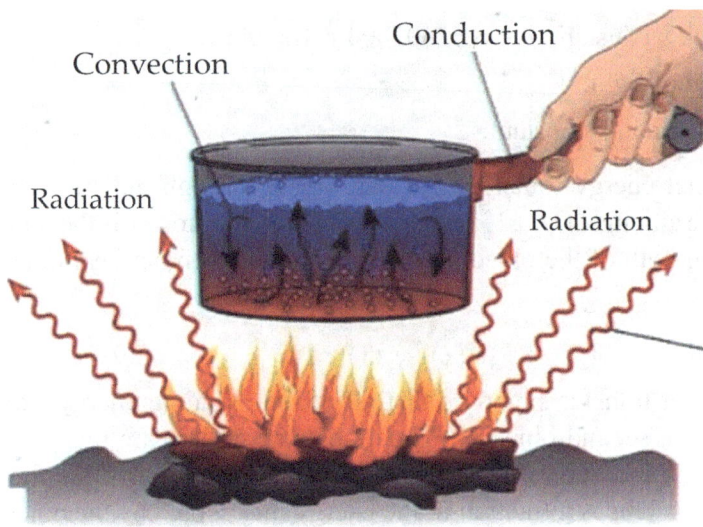

Figure 8.15. Heat transfer because of the combination of convection, conduction, and radiation.

Source: https://fl-pda.org/independent/courses/elementary/science/section4/4f12.htm.

REFERENCES

1. Arpaci, V. S., & Arpaci, V. S., (1966). *Conduction Heat Transfer* (Vol. 237, pp. 1–40). Reading, MA: Addison-Wesley.

2. Aziz, A., & Na, T. Y., (1984). *Perturbation Methods in Heat Transfer* (Vol. 1, No. 1, p. 212). Washington, DC, Hemisphere Publishing Corp.

3. Balaji, C., & Venkateshan, S. P., (1995). Combined conduction, convection and radiation in a slot. *International Journal of Heat and Fluid Flow*, *16*(2), 139–144.

4. Bég, O. A., Uddin, M., Rashidi, M. M., & Kavyani, N., (2014). Double-diffusive radiative magnetic mixed convective slip flow with Biot and Richardson number effects. *Journal of Engineering Thermophysics*, *23*(2), 79–97.

5. Bejan, A., & Kraus, A. D., (2003). *Heat Transfer Handbook* (Vol. 1, pp. 1–39). John Wiley & Sons.

6. Bejan, A., (2013). *Convection Heat Transfer* (Vol. 1, pp. 1–24). John Wiley & Sons.

7. Berber, S., Kwon, Y. K., & Tománek, D., (2000). Unusually high thermal conductivity of carbon nanotubes. *Physical Review Letters*, *8, 1–11.*

8. Biot, M. A., (1970). *Variational Principles in Heat Transfer: A Unified Lagrangian Analysis of Dissipative Phenomena* (Vol. 1, pp. 1–28). Biot (Ma) New York.

9. Boddington, T., Feng, C. G., & Gray, P., (1983). Thermal explosions, criticality and the disappearance of criticality in systems with distributed temperatures. Arbitrary Biot number and general reaction-rate laws. *Proceedings of the Royal Society of London: A Mathematical and Physical Sciences*, *390*(1799), 247–264.

10. Bourdin, E., Fauchais, P., & Boulos, M., (1983). Transient heat conduction under plasma conditions. *International Journal of Heat and Mass Transfer*, *26*(4), 567–582.

11. Carrigan, C. R., (1988). Biot number and thermos bottle effect: Implications for magma-chamber convection. *Geology*, *16*(9), 771–774.

12. Cess, R. D., (1966). The interaction of thermal radiation with free convection heat transfer. *International Journal of Heat and Mass Transfer*, *9*(11), 1269–1277.

13. Chai, J. C., Lee, H. S., & Patankar, S. V., (1994). Finite volume method for radiation heat transfer. *Journal of Thermophysics and Heat Transfer*, *8*(3), 419–425.

14. Chang, J. Y., & You, S. M., (1997). Boiling heat transfer phenomena from microporous and porous surfaces in saturated FC-72. *International Journal of Heat and Mass Transfer*, *40*(18), 4437–4447.

15. Chen, X. D., & Peng, X., (2005). Modified Biot number in the context of air drying of small moist porous objects. *Drying Technology*, *23*(1/2), 83–103.

16. Chen, X. D., Pirini, W., & Ozilgen, M., (2001). The reaction engineering approach to modeling drying of thin layer of pulped Kiwifruit flesh under conditions of small Biot numbers. *Chemical Engineering and Processing: Process Intensification*, *40*(4), 311–320.

17. Chung, T. J., & Kim, J. Y., (1984). Two-Dimensional, combined-mode heat transfer by conduction, convection, and radiation in emitting, absorbing, and scattering media-solution by finite elements. *Journal of Heat Transfer*, *106*(2), 448–452.

18. Dehghan, A. A., & Behnia, M., (1996). Combined natural convection-conduction and radiation heat transfer in a discretely heated open cavity. *Journal of Heat Transfer*, *118*(1), 56–64.

19. Dincer, I., & Hussain, M. M., (2004). Development of a new Biot number and lag factor correlation for drying applications. *International Journal of Heat and Mass Transfer*, *47*(4), 653–658.

20. Dombrovsky, L. A., (1996). *Radiation Heat Transfer in Disperse Systems* (Vol. 1, pp. 2–25). New York: Begell House.

21. Eckert, E. R. G., & Drake Jr, R. M., (1987). *Analysis of Heat and Mass Transfer* (Vol. 4, pp. 1–44).

22. Edwards, D. K., & Balakrishnan, A., (1973). Thermal radiation by combustion gases. *International Journal of Heat and Mass Transfer*, *16*(1), 25–40.

23. Fu, C. J., & Zhang, Z. M., (2006). Nanoscale radiation heat transfer for silicon at different doping levels. *International Journal of Heat and Mass Transfer*, *49*(9/10), 1703–1718.

24. Fujii, T., & Imura, H., (1972). Natural-convection heat transfer from a plate with arbitrary inclination. *International Journal of Heat and Mass Transfer*, *15*(4), 755–767.

25. Ganji, D. D., & Rajabi, A., (2006). Assessment of homotopy–

perturbation and perturbation methods in heat radiation equations. *International Communications in Heat and Mass Transfer*, *33*(3), 391–400.

26. Gee, D. L., & Webb, R. L., (1980). Forced convection heat transfer in helically rib-roughened tubes. *International Journal of Heat and Mass Transfer*, *23*(8), 1127–1136.

27. Hauf, W., & Grigull, U., (1970). Optical methods in heat transfer. In: *Advances in Heat Transfer* (Vol. 6, pp. 133–366). Elsevier.

28. Hayat, T., Khan, M. I., Farooq, M., Alsaedi, A., Waqas, M., & Yasmeen, T., (2016). Impact of Cattaneo–Christov heat flux model in flow of variable thermal conductivity fluid over a variable thicked surface. *International Journal of Heat and Mass Transfer*, *99*, 702–710.

29. Heijnen, J. J., & Van't Riet, K., (1984). Mass transfer, mixing and heat transfer phenomena in low viscosity bubble column reactors. *The Chemical Engineering Journal*, *28*(2), B21–B42.

30. Hensen, J. L., & Nakhi, A. E., (1994). Fourier and Biot numbers and the accuracy of conduction modeling. In: *Proceedings of BEP'94 Conference* (Vol. 1, pp. 247–256).

31. Howell, J. R., Menguc, M. P., & Siegel, R., (2015). *Thermal Radiation Heat Transfer* (Vol. 1, pp. 2–35). CRC press.

32. Hsiao, K. L., (2017). Combined electrical MHD heat transfer thermal extrusion system using Maxwell fluid with radiative and viscous dissipation effects. *Applied Thermal Engineering*, *112*, 1281–1288.

33. Huang, C. H., & Yeh, C. Y., (2002). An inverse problem in simultaneous estimating the Biot numbers of heat and moisture transfer for a porous material. *International Journal of Heat and Mass Transfer*, *45*(23), 4643–4653.

34. Incropera, F. P., Lavine, A. S., Bergman, T. L., & DeWitt, D. P., (2007). *Fundamentals of Heat and Mass Transfer* (Vol. 1, pp. 1–44). Wiley.

35. Jaeger, J. C., & Carslaw, H. S., (1959). *Conduction of Heat in Solids* (Vol. 1, pp. 1–24). Clarendon P.

36. Kakaç, S., & Pramuanjaroenkij, A., (2009). Review of convective heat transfer enhancement with nanofluids. *International Journal of Heat and Mass Transfer*, *52*(13/14), 3187–3196.

37. Kalb, C. E., & Seader, J. D., (1972). Heat and mass transfer phenomena for viscous flow in curved circular tubes. *International Journal of Heat and Mass Transfer*, *15*(4), 801–817.

38. Kirillov, I. R., Reed, C. B., Barleon, L., & Miyazaki, K., (1995). Present understanding of MHD and heat transfer phenomena for liquid metal blankets. *Fusion Engineering and Design, 27*, 553–569.

39. Kou, S., (1996). Transport phenomena and materials processing. In: Sindo, K., (ed.), *Transport Phenomena and Materials Processing* (Vol. 1, No. 1, p. 696). ISBN 0–471–07667–8, Wiley-VCH.

40. Kreith, F., & Black, W. Z., (1980). *Basic Heat Transfer* (Vol. 1, p. 259). New York: Harper & Row.

41. Kuehn, T. H., & Goldstein, R. J., (1976). Correlating equations for natural convection heat transfer between horizontal circular cylinders. *International Journal of Heat and Mass Transfer, 19*(10), 1127–1134.

42. Kurzweg, U. H., (1985). Enhanced heat conduction in fluids subjected to sinusoidal oscillations. *J. Heat Transfer, 107*(2), 459–462.

43. Lee, K. P., (1983). A simplistic model of cyclic heat transfer phenomena in closed spaces. In: *IECEC'83; Proceedings of the Eighteenth Intersociety Energy Conversion Engineering Conference* (Vol. 1, pp. 720–723).

44. Mezrhab, A., Bouali, H., Amaoui, H., & Bouzidi, M., (2006). Computation of combined natural-convection and radiation heat-transfer in a cavity having a square body at its center. *Applied Energy, 83*(9), 1004–1023.

45. Morgan, V. T., (1975). The overall convective heat transfer from smooth circular cylinders. In: *Advances in Heat Transfer* (Vol. 11, pp. 199–264). Elsevier.

46. Narayana, S., & Sato, Y., (2012). Heat flux manipulation with engineered thermal materials. *Physical Review Letters, 108*(21), 214303.

47. Özişik, M. N., Orlande, H. R., Colaço, M. J., & Cotta, R. M., (2017). *Finite Difference Methods in Heat Transfer* (Vol. 1, pp. 1–34). CRC Press.

48. Peng, X., & Wang, B. X., (1993). Forced convection and flow boiling heat transfer for liquid flowing through microchannels. *International Journal of Heat and Mass Transfer, 36*(14), 3421–3427.

49. Pham, Q. T., & Willix, J., (1990). Effect of Biot number and freezing rate on accuracy of some food freezing time prediction methods. *Journal of Food Science, 55*(5), 1429–1434.

50. Reddy, K. S., & Kumar, N. S., (2008). Combined laminar natural convection and surface radiation heat transfer in a modified cavity

receiver of solar parabolic dish. *International Journal of Thermal Sciences*, *47*(12), 1647–1657.

51. Rohsenow, W. M., Hartnett, J. P., & Ganic, E. N., (1985). *Handbook of Heat Transfer Fundamentals* (Vol. 1, No. 1, p. 1440). New York, McGraw-Hill Book Co.

52. Rousse, D. R., (2000). Numerical predictions of two-dimensional conduction, convection, and radiation heat transfer. I. Formulation. *International Journal of Thermal Sciences*, *39*(3), 315–331.

53. Sass, A., (1967). Simulation of heat-transfer phenomena in a rotary kiln. *Industrial and Engineering Chemistry Process Design and Development*, *6*(4), 532–535.

54. Sparrow, E. M., (2018). *Radiation Heat Transfer* (Vol. 1, pp. 2–25). Routledge.

55. Tan, Z., & Howell, J. R., (1991). Combined radiation and natural convection in a two-dimensional participating square medium. *International Journal of Heat and Mass Transfer*, *34*(3), 785–793.

56. VanSant, J. H., (1983). *Conduction Heat Transfer Solutions* (No. UCRL-52863-Rev. 1, Vol. 1, pp. 11–50). Lawrence Livermore National Lab., CA (USA).

57. Viskanta, R., & Mengüç, M. P., (1987). Radiation heat transfer in combustion systems. *Progress in Energy and Combustion Science*, *13*(2), 97–160.

58. Wang, B. X., & Peng, X. F., (1994). Experimental investigation on liquid forced-convection heat transfer through microchannels. *International Journal of Heat and Mass Transfer*, *37*, 73–82.

59. Xu, B., Li, P. W., & Chan, C. L., (2012). Extending the validity of lumped capacitance method for large Biot number in thermal storage application. *Solar Energy*, *86*(6), 1709–1724.

60. Xuan, Y., & Li, Q., (2000). Heat transfer enhancement of nanofluids. *International Journal of Heat and Fluid Flow*, *21*(1), 58–64.

61. Xuan, Y., & Li, Q., (2003). Investigation on convective heat transfer and flow features of nanofluids. *J. Heat Transfer*, *125*(1), 151–155.

62. Yoshida, H., Yun, J. H., Echigo, R., & Tomimura, T., (1990). Transient characteristics of combined conduction, convection and radiation heat transfer in porous media. *International Journal of Heat and Mass Transfer*, *33*(5), 847–857.

63. You, S. M., Kim, J. H., & Kim, K. H., (2003). Effect of nanoparticles on critical heat flux of water in pool boiling heat transfer. *Applied Physics Letters*, *83*(16), 3374–3376.

64. Zhu, D., & Miller, R. A., (2000). Thermal conductivity and elastic modulus evolution of thermal barrier coatings under high heat flux conditions. *Journal of Thermal Spray Technology*, *9*(2), 175–180.

65. Zhu, D., Miller, R. A., Nagaraj, B. A., & Bruce, R. W., (2001). Thermal conductivity of EB-PVD thermal barrier coatings evaluated by a steady-state laser heat flux technique. *Surface and Coatings Technology*, *138*(1), 1–8.

INDEX

Printed in the United States
By Bookmasters